ABOUT THE AUTHOR

Nick Brunacini lives is Phoenix, Ariz., with his wife, three daughters and two dogs. He has never met any of the American presidents although he did once have a drink with Andre the Giant in the San Diego Airport.

B-SHIFTER

BY *Nick Brunacini*

bshifter.com ☯ Phoenix, AZ

Front cover - Picture of author taken after a B Shift fire in 1988.
Photo by Bruce Varner

Back cover - The restored 1952 Mack L model pumper owned by Alan
Brunacini. This is the fire engine (Engine 1) that co-starred in chapter 1.
Photo by author.

Printed in the United States of America
Edited by Michelle Garrido

Second Printing

ISBN 978-0-9747534-6-1

Author's Note

This book is a memoir. I wrote it down the way I remember it. I have taken the liberty to change most of the names along with fire station and apparatus numbers. I hope you have as much fun reading it as I did writing it.

For my sweet bitches

Chapter 1
The Father

1966. Traffic was light, enabling Engine 1 to hurl down 7th Avenue at its top speed of 77 mph. The few pedestrians out that night heard the winding wail of the Federal Q mechanical siren over the inline 6-cylinder engine's roar before they caught sight of the rolling asylum. It was a 1952 Mack pumper with strong classic lines. The apparatus measured more than 28 feet long and weighed 15 tons, making it the largest, loudest, most serious red thing on the road.

Few objects represent the unrefined beauty, attitude and raw power of post-World War II America more than our 1950s-era fire engines. The truck wore enough chrome-plated pig iron to dress out half a dozen '50s era Cadillacs. The 10 feet of motor, hood, grille, bells, and sirens between the front bumper and the driver didn't do much for handling and maneuverability, but it conveyed a sense of purpose that today's cab-forward apparatus lack. Six gasoline-powered pistons the size of 1-gallon paint cans gave the beast profuse power and torque. The accelerating motor's growl generated a wall of sound and a palpable wave of pressure capable of silencing a fleet of Harley-Davidson motorcycles. Experienced engineers could coax the motor to blast a 10-foot column of flame from the rig's 4-inch exhaust pipe by synchronizing their shifts with the engine's RPMs. The experience of standing near-by while Engine 1 exploded down the road at full throttle terrified small children and left their parents breathless.

As the speeding engine approached the Thomas Road intersection, Dago Don, the engineer, came off the gas and double-clutched the square-toothed transmission from 4th to 3rd gear. Don tapped the brake pedal with his right foot while he stabbed his left foot down on the clutch, finessing the stick shift from 4th gear to neutral. He quickly let out on the clutch and drove it back into the floorboard while using his right foot to goose the gas pedal, raising the RPMs as he shifted from neutral into 3rd. The lower gear ratio combined with higher RPMs slowed down the truck. As the rig powered into the intersection, Don repeated the torque ballet, downshifting from 3rd to 2nd gear. He spun the large steering wheel a series of lefts while he put the accelerator to the floor. The motor was producing enough power to pull shade trees out of the ground. The large rear dual tires howled as they dug purchase into the hot asphalt for traction. As the open-cab engine came out of its turn, Don let the gigantic steering wheel loose, allowing the turn's built-up energy to put the front tires on a straight course down Thomas Road. The old engineer went from 2nd to 3rd gear and then into 4th, timing his shifts without use of the clutch. Several hundred feet out of the turn, the rig was back up to over 60 mph. The man was one with the engine.

The other occupant in the speeding engine's cab was a young Captain Alan Brunacini. He marveled over the skill and choreography the old engineer used to drive the red behemoth. Operating the apparatus required a combination of brutish power and finesse. Power steering was decades away from becoming standard equipment; steering the old truck required both an oversized steering wheel and Popeye forearms. The beauty of the large steering wheel struck the young captain. The absence of a roof had allowed the Arizona sun to bake the white plastic wheel. Over time, the wheel had developed small cracks and discolorations—giving it the appearance of ivory. Deep in Brunacini's heart, he felt the rig had more life-energy and soul than many people he knew. It was an odd thought considering the crew was currently rushing toward a burning building.

Engine 1 was housed in downtown Phoenix at Fire Station 1. The engine was staffed with six men (back in the day it was all men). The engineer and captain rode in the cab, while the four firefighters rode in a standing position on the tailboard. The fire chief was a fan of ordering apparatus without roofs. In the event a rig came in with a roof, the fire chief would have the shop cut it off prior to placing the apparatus in service. This layout permitted everyone to stand up while responding to calls. The only reason the engineer remained seated is because someone had to drive the truck. It was quite a sight to watch the four firefighters on the back step holding on for dear life while the captain stood up in the cab, surveying the horizon like an invading general.

Engine 1 raced toward the thermal column in the night sky. The young captain lay on the mechanical siren to move traffic out of their way. The firefighters riding the tailboard looked at the rising smoke and flame on the horizon while they finished their cigarettes. As the engine made its final approach toward the burning building, the crew got a full view of the spectacle before them.

Jack's Restaurant was your typical diner. It was a 100 feet long, 35 feet wide, single-story rectangle of a building. The fire had started in the kitchen at the back of the diner. It quickly extended into the attic through the vent system over the cooking appliances. The temperature in the back of the old diner was more than 1,000 degrees F. As Engine 1 made its final approach, the crew could see the four men of Ladder 10 feverishly chopping holes in the peaked roof to vent smoke and heat from the building's interior. They looked like lumberjacks wearing funny hats, possessed with a death wish as smoke and flame poured out around them.

Young Captain Brunacini ordered Dago Don to catch a hydrant on the way in. Dago slid the engine to a stop just past the fire hydrant that sat a few hundred feet east of the burning restaurant. The plugman stepped off the tailboard with the hydrant valve, wrench and the stretch of 2½" supply line it was attached to. When the plugman wrapped the hydrant with the supply line, Don buried the accelerator and headed for a spot in front of the diner. The three firefighters riding the tailboard paid close attention to the hose as it played out of the hose bed, parallel with their heads. The 50-foot sections of hose attached to one another with brass couplings. Getting hit in the head with 10 lbs. of brass traveling at 20–30 mph could instantly turn your average firefighter into a 3rd grader.

Engine 1's crew hadn't come to a complete stop as they watched one of Engine 10's firefighters use a pick-head axe to break out the diner's glass double doors. The heat from the kitchen fire was pushing smoke across the ceiling in the dining room. When it got to the front of the building, it had nowhere else to go. The laws of nature took over, and the smoke started banking down. Hot, black smoke poured out of the broken doors, but the fire in the back of the building continued to fill the top half of the restaurant with rolling black smoke. The magic of combustion was turning the solid fuel that made up the restaurant into flammable gas. Engine 1's crew had taken temporary refuge in their quiet places while they contemplated their journey into a place capable of dispensing violent death.

After Dago spotted the apparatus, one of the firefighters came off the tailboard with the scissor hose clamp and applied it to the dry supply line the plugman would soon charge with hydrant water. Captain Brunacini ordered his firefighters to pull a 2½" attack line to the restaurant's interior. The firefighters pulled 200 feet of attack line off the back of the truck, and after they got it into position, Dago Don charged it with 70 lbs. of pressure. The firefighters wrapped hose chains around the line, giving them the added leverage they would need to advance the line into the building. It's the same concept as using a harness to drag a pissed-off bull to a place it doesn't want to go. Once the nozzle was opened, the line's 1¼" smooth-bore tip would flow 400 gallons of water a minute.

In the 1960s, a firefighter's protective clothing consisted of a cotton-duck turnout jacket, a plastic fire helmet and thick work gloves. Their black denim Frisco jeans and street shoes served double duty as uniform items and firefighting gear. This protective ensemble is absolutely laughable when compared to today's protective gear.

Two of Engine 1's firefighters donned self-contained breathing apparatus (SCBA). The units weighed more than 50 lbs.; the steel air bottle that allowed the user to bring their own air supply into a lethal atmosphere accounted for most of the weight. Captain Brunacini threw on a cartridge-filter air mask. This unit weighed a lot less, but only because its main component was a paper filter. The canister mask was designed only to filter the bad stuff out of the atmosphere, much like breathing through a roll of toilet paper. The canister filtered out the big chunks of smoke, but fire gases and heat could pass directly into the user's respiratory system. Canister masks also had a tendency to clog with soot and large chunks of carbon particles. Firefighters always carried some type of hand tool in the pocket of their turnout jacket. When the filter clogged with large pieces of carcinogens, they used the hand tool to knock the chunks loose, enabling the wearer to breath again. This form of respiratory protection was eventually outlawed for firefighting. A decade or so later, it was resurrected by the people who design, produce and sell coffee filters.

Engine 1's crew made its charge through the front of the building, backing up Engine 10's attack line. Because the smoke was banked 3 to 4 feet off the floor, they had to stay low, advancing the line on their hands and knees. The crew spaced themselves out on the line, leaving a man at the door to hump hose inside while young Captain Brunacini and his two firefighters advanced their line under a canopy of pressurized black smoke. The deeper they went, the hotter it got. They didn't see the fire as much as they felt its heat and heard it vaporizing the building. The fire pops, roars and makes the building groan. It is a feast of consumption, giving the fire a significant presence that can completely destroy a building in less time than it takes to bake a pie. Engine 1's crew was in a race against the fire. In the next 5 minutes, the attacking crews would either kill it, or it would take complete possession of the building and kill any firefighter who couldn't get out.

Engine 1's crew broke through the smoke and heat after crawling 75 feet into the rear of the diner. As they entered the kitchen, the smoke lifted, restoring their vision. Now they had to contend with only heat, flame and a rapidly disintegrating structure. They opened the nozzle and went to work on the out-of-control blaze. The fire was now venting through the hole in the roof Ladder 10 had cut. This allowed Engine 1's crew to stand up, making it easier to play their nozzle back and forth across the burning back quarter of the diner. The interior attack crews were beginning to gain the upper hand when the chief arrived on scene and changed everything.

4

Battalion Chief Jimmy-Jack Hernandez arrived to the scene and took charge of the fire. His standard response to structure fires was to immediately put a defensive attack in place as quickly as possible. Once the defensive operation was up and running, Jimmy-Jack would switch to his second favorite management activity: directing traffic. Structure fires involve equal parts of excitement and danger, punctuated with bursts of great natural force. On the upside, they don't typically last much longer than half an hour. But a lot of high-intensity stuff takes place during those 30 minutes. Most fire departments devote at least one person (usually the highest ranking person on the scene) to take command of the event. This commander is responsible for managing all fire-department units, which eliminates the possibility of having a gap in the action. There is always someone in charge, so when a situation arises—like when firefighters inside the burning building have some type of life-threatening emergency, or when the building next door starts to catch on fire—the commander can react to the new situation and cause some type of prescriptive action to take place. The incident organization loses this capability when the incident commander takes it upon himself to venture out in the middle of the street with his fire-department flashlight and direct traffic. What the hell—it was the '60s after all.

That evening's defensive operation at Jack's Diner involved setting up elevated master streams to water bomb the fire from above. Jimmy-Jack ordered Ladders 10 and 4 to set up at opposing corners of the building. He had supply lines laid to both ladders. Each ladder would flow 1,000 gallons of water per minute from a height of 50 feet. The only problem with this tactic: Two attack crews were still operating inside the building.

Heat rises. This is one of the most basic laws of structural firefighting. This inescapable reality was the reason Engines 1 and 10 were making headway on the fire from interior positions. The fire was burning up and out of the hole in the roof. Given another few minutes, the interior crews could have extinguished the blaze. When Jimmy-Jack ordered the big water from up on high, it changed the "heat rises" dynamic. The 8 tons of water per minute pushed all the heat, smoke, flame and fire gases down on the two crews who remained inside. It was like being shelled with your own artillery.

Engine 10 had been the closest to the exit when the building's interior went dark and hot. Engine 1's crew was crawling for its life. After exiting the building, Engine 10's crew stayed by the door and helped three of Engine 1's firefighters out of the building. Jack Sullivan, Engine 1's nozzleman, was feeling his way down his attack line in the absolute darkness and ear-blistering heat his chief had just rained down upon him. He was 40 feet from the door when he came across a large object lying across his line. He quickly figured out it was his unconscious captain.

The young Alan Brunacini was 5'8" and weighed 250 lbs. When he was alive and properly oxygenated, he could bench press 400 lbs. Jack Sullivan was 5'10" and weighed 175 lbs. The eternity Sullivan spent pulling his dead captain toward the doorway he thought to himself, "Why the hell couldn't he be pulling my ass out of here?"

Soon after Chief Jimmy-Jack ordered the aerial water attack, the massive smoke and heat inversion burned the bottom half of Brunacini's cartridge filter and clogged the remaining filter with carbon, soot, heat and dozens of exotic carcinogens. His next few breaths mainlined smoke and carbon monoxide into his bloodstream. This put his lights out.

One of the first groups of people Sullivan saw after he got his dead captain safely outside the doomed diner was an Associated Ambulance crew. In the mid-'60s, the fire department didn't deliver EMS. It was a fortunate happenstance that an ambulance crew was standing by when Sullivan dragged Brunacini out of the building. Whether it was fate or just dumb luck, the ambulance crew quickly determined the young captain wasn't breathing. They stuffed an airway down his throat, pumped pure oxygen into his lungs, transported him to the hospital and hoped for the best. By the time the ambulance pulled into the hospital ER, Captain Brunacini woke up with an aching brain full of carbon monoxide, wondering where the hell he was.

I've often wondered where young Captain Brunacini went when he died for those few minutes. Did he recall all the details of his life, or was his death image a rerun of his life's most vivid memory? Maybe he relived the trip he took to the grocery store with his mother when he was 6 years old. Young Al and his mother were running errands in downtown Albuquerque. He was sitting next to his mom in the front of their 1944 Chevy when a fire engine turned the corner in front of them. A building up the street was on fire. Mrs. Brunacini can best be described as a stern woman. The small kindness of pulling the car over and allowing her young son to stand up in the front seat and watch the spectacle of the fire forever changed that little boy and the industry he became a part of. The young Alan V. Brunacini watched in amazement as the members of the Albuquerque Fire Department attacked the burning structure. The emotional impact of that firefight, coupled with Al's driven nature, put the man's destiny in motion. That day, Alan Brunacini joined an elite club of 6-year-olds who make the unwavering decision concerning what they will do for their entire professional lives. Twenty-five years later, Alan Brunacini would die on duty for a couple of minutes, wake up and spend the next 40 years becoming the greatest fire chief since Benjamin Franklin created the American Fire Service in 1736. I don't know if this is the trip he took when he went away for a little while in that burning diner because he just shrugs when asked about it.

Al had joined the Phoenix Fire Department in June of 1958. He drove his 1955 Chevy BelAir to the city personnel office and filled out an application. A few days later, he showed up to the fire department drill tower and ran through the physical-agility course. After he successfully finished the agility test, one of the training officers and a chief interviewed him. He was hired that day. The next morning, he showed up to the training facility where he met the five other guys in his training group. Their academy lasted a whopping three days.

"The small hose is used to put out small fires; we use the bigger hose to put out big fires. Over there are the ladders. We use those to go up."

Three days later, his recruit class joined the other 200 members of the department that protected 100,000 people residing in the 200-square-mile city.

Big Al joined a group of men who were mostly veterans of World War II. These guys had fought the fascists and won. After leaving the military, they joined—and in some cases went back to—jobs within the fire department. The fire department looked and ran a lot more like the military back then than it does today. Life was simple. You did what you were told, worked hard, paid attention, kept your shoes shined, and if you were fortunate, you promoted. But the most important hallmark of an up-and-coming officer was not to change anything. We hate change. Keep the mission simple: water good, fire bad. Big Al tells stories about being one generation removed from officers who were still pissed off that Dolly and Clara-Belle (the department's last two horses) were put out to pasture in favor of motor-powered apparatus. In their eyes, 40 years of progress wasn't enough to overcome the sin of change.

This organizational philosophy kept the world small and easy to manage. It had been just over a decade since the end of the last Great War, and the Phoenix Fire Department (just like every other fire department) was a man's world governed by simple rules. Respond on fire calls, and when you get back to the station, wash the tires on the rig before you pull it into the apparatus bay. (This was a holdover from when horses pulled the rigs. If you didn't wash the horse shit off the wheels, the station would stink. In our infinite wisdom and hatred of change, we were still doing it—40 years after the last horse left the building.) But the most important order of business was not to piss off the chief. The department's focus was crystal clear. Our mission, back then, was to put out fires, perform the occasional building inspection and license kids' bicycles. Anything that fell outside these three service menus was some other group's responsibility. We were charged with keeping the city from burning down and making it easier to find stolen bikes. Performance was judged on the cleanliness of the station and equipment, the length of your hair, how well you followed the volumes of rules and regulations, and little else. Our organization mirrored the American society of the late '50s.

Big Al quickly moved up the ranks. He started his career riding Engine 1 as a firefighter. After spending five years as a firefighter on Engine 1, Big Al finished at the top of the engineer's list and drove Engine 1 for a couple of years. Quickly tiring of operating his beloved talisman, he repeated this process with the captain's test and ended up running Engine 1. A few years later, he completed the trifecta and became a battalion chief in charge of the 1st Battalion. One day, his wife accused him of loving Engine 1 more than he loved her. He smiled at her and replied, "Dear, while that may or may not be true, I will always love you more than I love Ladder 1."

Within 10 years, he had promoted from a firefighter, riding on the tailboard of Engine 1, all the way to battalion chief, supervising half a dozen captains, all old-timers who had a hand in training him. Now he was their boss. Big Al was a young Turk who had big plans in store for his beloved department. He and the other young rabble-rousers were pulling the organization in a brand new direction. We were going to jump into the EMS business. This didn't set too well with the large majority of the older crowd. They believed if somebody needed doctoring, you took them to the hospital. The thought of poking people with needles, shocking their hearts with complex and foreign-looking devices, and injecting them with powerful drugs was too much for a lot of them to fathom. They couldn't imagine themselves doing this, let alone all these new kids the department had been hiring. Many of the old-timers thought the baby-boomer generation didn't know their ass from a hole in the ground or the value of a nickel. Others felt EMS work was demeaning. We aren't nurses; we are firefighters goddamn it, heroes who rush into places where angels fear to tread. The new chief was going to make everyone learn about viruses and bandaging wounds. This was clearly women's work and certainly did not belong in the testosterone world of intrepid fire gods. A larger segment of our fire department embraced the new service. It was the early 1970s, after all — a very far-out time.

From the time Alan Brunacini was 6 years old, he managed his life around becoming the fire chief of the Phoenix Fire Department. He saw the value of education and the necessity of having a college degree. One of the first things he did when he got hired was request an educational leave of absence. He wanted to go to Oklahoma State to get a bachelor's degree in fire science. This was one of the only universities in the country, or world for that matter, that had a fire-science program. The fire chief granted his request, but the fact that an upstart 21-year-old firefighter was leaving town to learn something raised quite a few eyebrows within the establishment. Many of the department members felt that if you didn't learn it from them, it wasn't worth knowing.

Martin, one of the older captains on our department, put Big Al's educational pilgrimage into perspective for me. The two of us had been talking about the different schooling options a firefighter had at the time. It was the early 1990s and colleges and universities were starting to add more graduate tracks for the upwardly mobile fire professional. He told me the majority of our members took the plethora of education options that they had for granted.

"All of these kids coming on today have the attitude that college is their right," Martin said. "It hasn't always been that way. It all started with Alan. He was the first one who pursued a degree after he was hired, and he was on probation when he did it. It took balls to go into the chief's office and ask for an educational leave."

He must not have liked the way I looked at him. Before he continued his history lesson, he lit a cigarette and blew a cloud of smoke in my direction.

"When I took my first oral interview to get on this job in the early '60s, I had a bachelor's degree. I was proud of this, so I made mention of it to the oral board. Everything seemed to have been going pretty well up to that point. The chief on my board was a big, good-looking guy, the kind of man women give themselves to. After I mentioned that I had a degree, he looked at me and said, 'Spell Khrushchev for us.' This request confused me. 'I don't understand the question sir,' I said. He replied, 'A college-educated, left-wing faggot such as yourself surely ought to be able to spell the name of the largest Communist country's leader, son. Spell the fucking name.'

"I apologized for wasting their time and got up and left. The next time I took the test, I didn't get any more advanced than the workings of a monkey wrench during my interview. What I went through was mild compared to the climate Chief Brunacini had to deal with. Like I said, today's generation doesn't know jack shit about this organization's past and how well they have it today."

Martin fell back into his story. "Come to think of it, my boy, our chief and Napoleon had a lot common. They were both gifted by timing and fate. Never underestimate the importance of being in the right place at the right time."

During the first two decades of his career, many of the things Big Al observed and learned from his former bosses can be filed under "what not to do." Back then, the discipline process wasn't designed to correct problems; it was intended to punish anyone who upset the boss, took any action without prior approval, or caused a previously clean object to become dirty. Back in those days, the bosses used a management model developed by European royalty in conjunction with the Vatican during the heady days of the inquisitions. Big Al did not like this management style and came to realize it was bad for the department. The members would never be able to outperform the organization.

One of Alan Brunacini's basic management tenets is that rules should support the people who actually get the work done. The young upstart Brunacini reasoned the workforce would mirror the treatment they received at the hands of their masters onto the customer base. (I believe Dr. Phil refers to this phenomenon as "transference.") At a young age, Al felt most of the department's rules and discipline processes got in the way of the organization's core mission of delivering emergency service.

One night, when Big Al was the battalion chief in downtown Phoenix, he was returning to his quarters at Station 1. He had slowed down to turn into the station when he noticed something strange. The station was completely dark, as if the power had been shut off. The only light emulating from the station was from a makeshift movie screen the crews had created by hanging a sheet over the front window. Images of naked people doing the nasty were bleeding through the sheet. It was like a drive-in movie for street people. (This was the technology of the day. The men had to enjoy their "art films" with noisy and unreliable 8mm projectors. In these vintage films, all of the actors had hair "down there.")

Big Al pulled his rig around back and went inside. He found all 20 men assigned to the station assembled in the rec room. They were following tribal seating arrangements—older men closest to the activities, with back-row seating reserved for the very young. Al sat down next to the most junior guy. This youngster had just finished his one-month academy and was following the rituals of his new clan. It took several minutes for the new guy to notice his boss's boss sitting next to him. I'm sure the kid was a decent person and going through some type of Catholic- addled guilt, plus a dose of mild homophobia about having an erection while seated next to his battalion chief. The young member looked very uncomfortable and worried about the chief catching them doing something "immoral." He started to get out of his chair when Big Al grabbed him by the wrist and quietly but firmly told him, "Sit down, now." The firefighter did as ordered. As he slumped back into his chair, all the life drained out of him and his job passed before his eyes. Big Al leaned over, pointed at the screen and whispered, "You were going to miss the best part."

On public porno night, Battalion Chief Brunacini quietly took the officers off to the side and told them it wasn't very bright to host public showings of the human mating ritual and told them to be more discreet in the future. No one was executed, flayed, suspended or written up.

In 1978, Alan became the chief of the department. He quickly developed a two-prong approach to managing the Phoenix Fire Department. The two driving forces were customer service and firefighter safety. This visionary (yet elegant and simple) organizational philosophy turned the Phoenix Fire Department into one of the premier organizations in the world. In 1997, the "Wall Street Journal" proclaimed, "Brunacini has created a customer-service culture that few private companies could match."

During his 28-year reign, Chief Brunacini ushered in paramedic engine companies, and he developed and refined a set of hazard-zone management procedures used by fire departments all around the world. He is universally regarded as the biggest and most respected advocate for firefighter safety. Hundreds of firefighters are alive today as a result of the safety initiatives he has championed. He commented that the fire service spends more money maintaining its apparatus than in taking care of its firefighters. He was able to persuade (or cajole) city government into building and staffing the Phoenix Fire Department's own Health Center. To date, it has saved the lives of more than half a dozen firefighters through yearly physical exams.

Alan Brunacini developed and preaches about the concept of fire-department customer service and taking care of "Mrs. Smith." During Big Al's fire-chief career, more than 40 chief officers from the Phoenix Fire Department went on to become fire chiefs in other cities. He is regarded around the country as "America's Fire Chief." Union leaders hailed him as a visionary for the labor-management process he invented and implemented for the Phoenix Fire Department. He was the driving force behind getting the 20-plus fire departments that protect Maricopa County (an area the size of Connecticut with a population of more than 3 million people) to automatically respond and operate as a single department. In the end, his biggest kudos comes from Phoenix firefighters who refer to him as the father (and sometimes the mother) of the department.

In a very real sense, this book would have never been penned if not for Alan Brunacini. In 1959, Alan and his wife, Rita, sired an offspring that would follow in his father's footsteps. That would be me.

Chapter 2
The Sons

When I was 6 years old, I wanted to be a Chevy truck. I have been imbalanced most of my life, but this is not entirely my fault. I was forcibly pulled from my mother's womb with a set of stainless-steel salad tongs. Adding insult to injury, the doctor proceeded to smack me hard on the ass, and the very next day, some sadistic bastard cut off part of my dick. This series of events was so traumatic, I wasn't able to walk for a year. I have spent my life coping with these weighty issues.

I eventually abandoned my earliest childhood career dream and followed in my father's footsteps. From a very young age, my father knew who he was and what he was going to do with his life. For any normal person, it would be daunting to join the fire service as Alan Brunacini's son. Imagine if the fruit of Babe Ruth's loins decided to pursue a career in baseball. It's not difficult to imagine drunken fans threatening Bambino Junior's life while pelting him with half-eaten hot dogs and cups full of warm beer. When a mob verbalizes their expectations, it can be hurtful, but I can honestly say this has never been an issue for me.

Most human beings spend more than a third of their adult lives working. What we do for a living becomes one of our most defining attributes. In fact, there are a handful of careers that instantly define a person. Mother, circus performer, florist, third-world dictator, porn star and firefighter fall into this category.

My early fire-service education began when I was a small child. Most of my dad's friends were firefighters, so it comes as no surprise that my godfather was one. Uncle Sarkis joined the fire department in the late 1950s. The man had a severe obsessive-compulsive disorder and was perfectly suited to fill the role of Dutch uncle. Sarkis had no children of his own, so he jumped at the chance to become my Catholic-approved man guide.

Sarkis was of Armenian heritage. He was a can of tomatoes short of 6 feet tall and maintained a lean, hard 175-lb. frame. His black, slicked-backed hair was in the early phases of male pattern baldness. He had small, dark, deep-set eyes, and his teeth looked liked he had filed them to form a perfectly straight tooth line. He didn't walk like normal people; it was more like he threw himself from place to place. During my 20 or so years of sporadic contact with the man, I never knew him to have a spot of dirt on his clothing or a hair out of place. He always looked like a brand new deck of cards. For the most part, he did not use the restrained tone of normal speaking, he used a drill sergeant's voice. Most of his conversation was serious and delivered as orders. He didn't ask many questions. The older I got, the more certain I became he was clinically insane.

My parents have pictures documenting my blessed baptism into the fun-filled Catholic Church. My young and fresh parents are standing next to an older priest wearing what looks like a very gaudy Mexican fiesta muu-muu. Uncle Sarkis is standing on the other side of this very solemn agent of our Lord. I appear as a freshly circumcised infant wearing a frilly pillowcase. The snapshots follow the progression of events: my radiant mother and homicidally bored father looking on as the priest holds me; the priest dunking me in Jesus' very own bucket; the priest passing me to Crazy Sarkis. A picture taken in the church parking lot sums up the full implications of the morning's activities. A glaring Uncle Sarkis is holding me in the cradle of his left arm. He is wearing a Western-style gun belt and pointing a large revolver at the camera. Any communist who dared get within shooting range of his newly anointed godson would be dead before he hit the ground. God bless Jesus, Mr. Colt and the U.S. of A.

The first vivid memory I have of my rabid uncle is from Christmas Eve when I was 6 or 7 years old. Our nuclear family was hanging out at home. Dad was doing homework for his master's degree, and mom was baking Christmas treats for the family unit. We kids were watching one of the five available channels on our trusty 22-inch, black-and-white TV. Color television had yet to reach the masses, and cable hadn't been invented yet. Pork Chop, our Bullmastiff, lay snoring and farting simultaneously in front of the Naugahyde sofa my brother, sister and I were perched upon. The three of us were enthralled as we watched an animated snowman speak with Burl Ives' voice. It was a scene the church would approve of. The serenity of our evening was interrupted by a knock at the door. I trotted across our living room to answer it. (This event took place during the mid-1960s—a time when young children were permitted to answer knocks at their front door.)

When I opened the door, my lunatic uncle—dressed as Santa Claus with the big white beard and all—greeted me. I was no detective at this young age and was completely clueless as to who was standing before me. My confusion quickly turned to fear when the imposter started barking "Ho! Ho! Ho!" with all the Christmas cheer of an Arab slave trader. As I tried to conjugate what was happening, Pork Chop knocked me down and lunged at the threat. Uncle Sarkis had lightning reflexes from years of quick-draw practice. Before Pork Chop's quickly closing jaws found flesh, my uncle

14

sprang out of the way so the large dog's teeth tore into Santa's big, red bag of gifts. During the course of the brief wrestling match, my brother and sister began screaming at the tops of their lungs, "Pork Chop's killing Santa!" I was curled up in a ball under our 150-lb. dog and the fake Kris Kringle. Pork Chop had a death grip on the bag and was viciously whipping his head back and forth with animal power and ferocity. Sarkis knew the bag was the only thing protecting him from serious pain. He was not about to let it go. My father came around the corner as Pork Chop shook my uncle into an end table, sending both the table and Sarkis careening across the room. Dad grabbed the dog by its collar, but Pork Chop wanted a piece of Santa's ass and continued lunging and gnashing his teeth. My uncle was wedged into a corner. He had his right hand in his boot. My first thought was Pork Chop had bit a chunk out of Santa's leg. This theory was quickly dispelled when my uncle's right hand produced a small revolver. He pointed it at Pork Chop and screamed, "Alan, that's an excellent animal! It was just protecting the family! I don't want to shoot him, but I will defend myself!"

As my powerful father scooped up our crazed dog into his arms, he looked at my uncle and told him, "It's Christmas, Sarkis. Why don't you put that fucking thing away." I do not recall any more of that evening's activities.

☠☠☠☠☠☠☠☠

My father is not an outdoorsman. He never has been, and I seriously doubt he will ever be. Any outside time my brother and I spent with our dad involved shovels and other ancient human-powered tools. Growing up, "outside" meant building something. When we ran out of space to build, we tore something down so we could build something else.

I think my father was happy Uncle Sarkis was part of our lives because he provided his sons the opportunity to do the other boy stuff my father couldn't stand doing. Uncle Sarkis loved guns and hunting. When I was a boy between the ages of 9 and 13, I would go on outings with my uncle and his hunting buddies. As I reflect back on these happy times, they strike me as training and preparedness missions for a paramilitary group. We did hunt and shoot guns, but we also buried guns, ammo, food and water out in the middle of the desert. Don't get me wrong: Even though my uncle was a bit off balance, as were many of his hunting associates, he was fundamentally a nice person. When we ran across other hunters, everyone was polite and helpful. For example, we would always stop and assist people who had broken down on the side of the road. It seemed to me the thing that bound them together was they all felt the shit was going to hit the fan on a global scale, and they would be ready when it did. Years later, I realize I had been safer with these survivalists than I would have been in a Catholic Sunday school.

One day, I was at my uncle's house after one of our hunting expeditions. We had cleaned the guns and had taken the birds we shot-gunned out of the sky inside for Sarkis' wife to cook. At that point in Sarkis' life, he was married to a German woman named Grisilda. She was short and thin and spoke with a very heavy accent. She had been born in the fatherland and came to the United States with her mother after Hitler and his young bride took their gasoline shampoo.

Sarkis and Grisilda had two German shepherds, Fang and Sally. Grisilda heaped them with affection, while my uncle trained them to go for the soft tissues of the groin and throat. My uncle used to dream out loud about the day some drug-addicted rapist/burglar would break into his house and meet his doggies. We were in the garage after putting away the last of the hunting gear when my uncle raised the garage door so he could park his truck. Both dogs were lying on the garage floor when the mailman appeared at the end of the driveway. Apparently, the dogs could not differentiate between drug-crazed burglar/rapists and postal carriers. Both dogs came off the floor, barking and snarling with bad intentions. They were quickly closing the distance that separated them from the mailman. Sarkis was standing at the end of the garage when he calmly reached over and removed a coiled bull-whip from the wall. He held the handle with one hand and casually flicked the whip out to its full length of 8 feet. He then started working the whip, coming within inches of the dogs' noses. The mailman stood frozen midway up the driveway. Uncle Sarkis was slowly backing toward the mailman, keeping the whip going while facing his two prodigies. He was almost out of the garage when Grisilda opened the door that led from the house into the garage. This disrupted the rhythm of my uncle's whip strokes and his backward pace toward the mailman.

Grisilda shouted, "What the hell am I supposed to do with these dead birds?!" before she realized she had placed herself between my uncle and the two angry dogs. As the whip cracked inches from her face, my uncle screamed, "Get back into the house, you Nazi bitch!" Grisilda started screaming at the dogs in German, "Mien hitchen gritchen strudel blah, blah, blah!" My Uncle started whipping faster and screaming at Grisilda, "English goddamn it! I don't want you confusing the dogs with your bastard native language." Grisilda replied something in German that neither the mailman nor I understood. Whatever it was, Sarkis understood and let loose with a tirade about her father deserving what the Russians had done to him. Grisilda ran into the house crying.

The mailman had had enough. He dropped the mail to the ground and ran. My uncle continued cracking the whip at the two lunging dogs and screamed at the fleeing mailman that he was a coward.

☠☠☠☠☠☠☠☠

My fire-service education continued at the age of 14 when I went to work for a firefighter friend of my father who owned a restaurant/bar called Oscar's Casa de Taco. The restaurant was located in the northeast section of Phoenix, several hundred feet north of the firefighter's union hall. This supplied the bar with a constant customer base.

Entering the workforce at such a young age necessitated being dropped off and picked up for work. One Saturday after 1 a.m., I was sitting on a planter box built out of railroad ties, waiting for my ride. As I sat on the uncomfortable perch, half a dozen senior members of the Phoenix Fire Department poured through the bar's exit. Before getting in their cars, the group shared a six-pack of beer, enjoying a final drink before embarking on their drunken treks home. I sat off to the side, completely hidden from the group as they began arguing about World War II naval battles. Within seconds, half-full beer cans were being thrown, and the group climbed into their respective vehicles while calling one another bad names. Gravel sprayed the parking lot as the cars fishtailed out of control. As fate would have it, one car bumped into another, and seconds later a full-fledged demolition derby ensued. Within minutes, all of the participants were making their exit in a giant cloud of dust and hissing radiators. One of the drivers was already several hundred feet up the road when his brake lights flashed red and his car skidded to a stop. He sped backward toward the parking lot. We both disappeared in a cloud of brown dust as he slid his car to a stop next to me. When I could see again, I was eye-to-eye with a 45-year-old man who looked like he was trying to recall long stretches of his life. He came out of his alcohol-fueled fog, smiled at me and asked, "You're Bruno's kid, aren't you?" Knowing a right answer didn't exist, I replied, "Yes." He appeared to struggle with maintaining consciousness for a few beats before telling me, "Your old man's alright. He pisses off the chief. Our chief is an ironclad fuck-stick. You need a ride home, kid?"

Before I could answer, one of the departed demolition derby participants appeared out of nowhere and rear-ended him. The two of them drove north to a calliope of racing motors and scraping metal.

I worked at Oscar's for the next decade, promoting to the rank of bartender at the illegal age of 17 and holding this coveted position for six years. This career refined my knowledge regarding the social requirements and group dynamics of the fire-service cult I would eventually join. A wide range of generations hung out at Oscar's, everyone from old alcoholics to young guys trying earnestly to become old alcoholics. The crowd included our department's most militant union goons, rank-and-file firefighters, paramedics, captains and high-ranking chiefs. Three years of pouring liquor better prepared me for a career in the fire service than graduate school.

One evening, the bar was packed with patrons. Several men wearing camouflaged Army uniforms occupied the corner of the bar; they were generally being a nuisance. Frank, their ringleader, was a colonel in the Army Reserves and a captain on the fire department. His group had been drinking since lunch. Oscar, the proprietor of the establishment, was sitting next to Colonel Frank. I had just finished making a couple of pitchers of VO and water for the A Shift softball team. (I believe we were the only saloon in town that sold gallon pitchers of whiskey and water.) I went over to ask Oscar if Frank and his natural-born killers needed their drinks refreshed. He had me set up another round on the house. Colonel Frank took his bourbon and Coke, threw his head back and proceeded to drain his glass. As he leaned back to get every last drop, he went too far and took a 4-foot fall flat on his back. When he didn't move, I assumed he had either passed out from overconsumption or knocked himself unconscious. Oscar looked down at the comatose soldier. "That's what I like about Frank," he said. "He always knows when he's had enough." After a couple minutes, Frank still hadn't moved, and his body was creating an obstacle. I used a stream of delicious ginger ale, delivered from the soda gun, to spray his drunk ass back into consciousness. It got Frank up off the floor and inspired the crowd to tip generously.

☠☠☠☠☠☠☠☠

I joined the Phoenix Fire Department on Feb. 12, 1980. There were 16 of us in my academy class. We all had competed with the mass of humanity that signed up to take the entrance exam. The selection process started with a written test. After scoring well enough on the written exam, we had to

18

pass a physical-agility test, which included a 12-minute jog. (In the past 28 years, I've yet to get to the scene of any call and jog for 12 minutes.) After we passed the agility test, our two scores were totaled and we were ranked on an eligibility list. About a month prior to our academy's start date, the fire department invited 160 candidates to interview for the 16 firefighter positions. More than 2,000 people signed up to take the firefighter's test in 1979, and about 75 of us were hired and trained in three academy classes during the one-year life of the eligibility list. I was a member of the third and final class. On the first day of our 12-week academy, I met the other 15 lucky grand prize-winners. This was the first step in preparing us to join the other 800 members of the department, protecting a city of almost 1 million people that encroached upon 400 square miles of the Sonoran Desert.

Even though they occurred decades apart, the assimilation processes my father and I endured had a lot in common. We were both trained by experienced members, who happened to be older men. The apparatus and equipment were pretty much the same. In fact, most of the firefighting tools I learned to use in the training academy were not much different from the ones firefighters used hundreds of years ago when horses pulled the apparatus. The tactics used to extinguish structure fires are basically unchanged from those used in the 1700s. The organizational model used to manage the department is also more than 200 years old. The fire chief sits at the top of the food chain. Everyone else falls under him (even in the late 1970s all fire chiefs were men). Efficiency experts like to call it a "paramilitary model." Our training officers described our place in the organization in very eloquent terms: "All of you are lower than whale shit." I'm sure my father received the same analysis about his organizational position from his training officers.

I was discussing how little things have changed with my father one day when he told me a story about the history of the American fire service. "Our founding fathers started the infrastructure for our society. They formed the military, started post offices, built roads and bridges, and formed our government. Benjamin Franklin and George Washington are both prime examples. They were the fire chiefs in their communities. One day, Washington was preparing to take a brief trip from his Alexandria, Va., home. Before he left, he told his fire company not to change anything until he got back. He died on his trip. The fire department is still following his orders and hasn't changed a thing. I'm sure some of them are still waiting for him to return. Son, that's the American fire service."

Mornings during my three-month academy were spent on the grinder. The grinder was a 1-acre concrete slab that contained two concrete buildings, a two-story smokehouse, a six-story drill tower and four fire hydrants. I never found out why it was called the grinder. We never ground anything out there, and it never ground any of us. A more appropriate term would have been the incinerator. Most of us received a collage of first- and second-degree burns on our ears, faces and hands over the course of our training. During our morning sessions, we learned how to do the firefighting part of our new jobs. We spent our afternoons learning how to do the newest part of our jobs—providing emergency medical services (EMS) to the community.

When my father began his career with the fire department, paramedics hadn't been invented yet, and prehospital care really didn't exist. At the time, EMS was delivered by one of two entities. Private ambulance companies were used to get patients from the scene to the hospital and, beyond administering oxygen, provided very little in the way of treatment. Private ambulance services were very sparse and, in many cases, had the same response time as a pizza delivery. The other entity that occasionally responded to medical calls was the mortician. If the patient was alive, he took them to the hospital. If they were dead, he took them to his funeral home. The financial return was always higher if the patient had died. I don't think I want the guy charged with saving my life to have that kind of conflict of interest.

Prehospital care changed in a big way during the early 1970s. During the Vietnam War, the Army found that if they could treat wounded soldiers in the field, their survival rates skyrocketed. Some forward-thinking social scientist theorized there wasn't much difference in treating a gunshot wound in a hostile foreign land than there was in treating one at home; the key to the entire system was to get early treatment. Fire departments were already set up to respond to most places within their city in 4 minutes or less. We were a natural place to try out the new service. California was one of the first places firefighters became paramedics. This new service-delivery program became so successful that Hollywood made the TV show "EMERGENCY!" Every week, millions of people got to watch paramedics save the usual five or six victims from fires, heart attacks, auto accidents and snakebites. It wasn't long before the viewing public started demanding paramedics in their communities.

Our afternoons spent learning EMS were dry and boring. Our firefighting training was fun and exciting. One of my fondest academy memories is of the time I watched both of my training captains get blown out of the two-story smokehouse. Our training officers used to hide dummies in there, then ignite wood pallets, mattresses, tarpaper or the occasional car tire. When they deemed the inside of the structure lethal enough, they sent us in to find the hidden dummies and extinguish the inferno. On more than one occasion, we pulled still-burning dummies out of their fiery home.

One morning, our instructors had prepared the smokehouse for our final evaluations by loading the interior with flammable goodies and spreading quite a bit of gasoline around so the fire would be extra hot and visually exciting. Their routine was to work from the most interior point toward the exterior exits. Both instructors were standing in their respective doorways—one on the first floor, the other on the second—when they lit their flares and threw them back into the explosive atmosphere they had just created. If they were to do this today, OSHA or any other of a dozen different regulatory agencies would show up and take them to jail.

Nothing happened for a full minute. The captains looked at one another and shrugged. Three of us sat on an idling fire engine, waiting for one of the training captains to order a certain hose lay and attack point when they were happy with the size of the fire. The rest of the class hung out under a cluster of shade trees, watching and waiting for their turn. Both captains were wearing only fatigue uniforms when one of them decided to go back inside and find out why flame wasn't blowing out of the building's windows and doors.

All fuel needs a certain amount of oxygen to initiate and sustain the combustion process, but they had loaded the interior so full of gasoline that the air/fuel ratio was too rich to burn. The gas fumes had displaced most of the air. When our captain pulled open the door, he introduced enough outside air to even things out. He had just taken a step inside the door when the entire interior lit up. The power of the unburned vapors igniting at once had a pressurizing effect that began deep inside the building then blasted its way to the exterior. Our captain was blown out of the building and into a railing that kept him from making a two-story drop. He finished his combustion-propelled ballet by rolling down a set of exterior stairs. When he stood up on his wobbly legs, he was smoldering. Our other captain was blown on his rear-end, but for the most part he had escaped the kiss of fire. The entire event took less than 5 seconds to play out, but it took at least that long for most of our group to figure out what had happened and for several of us to break into riotous laughter. After both captains ascertained they hadn't been burned to death, their attention turned to the laughing young neophytes. We spent the remainder of the day inside the burning bunker. We never had this type of fun during EMT training.

☠☠☠☠☠☠☠☠

A week or so after my class graduated from the training academy, we were invited to attend our first union meeting. The American fire service is represented by a very active labor organization. The vast majority of paid, career departments are members of the International Association of Fire Fighters. That evening's union festivities included voting the newest set of firefighters into the fraternal brotherhood. This process was nothing more than a formality; we had done all the paperwork to join while we were still in the academy and no one had ever been vetoed from joining. After all, a dues-paying member is a dues-paying member.

The union hall was packed. In previous years, the relationship between our local and city hall had become so dysfunctional that talks of a strike vote loomed large. The combination of free beer and a planned showdown with the city masters had the crowd jacked up. After the newest group of firefighters was formally voted in, the group discussed what action to take concerning contract talks with the city. The union leadership said going on strike had too big a downside. They explained we weren't mad at the citizens, we were upset with the mayor and council. The packed meeting quickly spun out of control. Things became hushed when a big Swede standing in the back of the room began a nonsensical, impassioned tirade.

"We have got to get the newspapers on our side. Whoever controls what people read everyday controls what this city does. Baby, I think that is the key to what we're looking for, that kinda control! It's what the man wants! We think we can do that, then we need that want, and to be able to control it. If we can do that then we can do it." As the Swede got into the full swing of his tirade, he began shouting at the top of his lungs. By the time he finished, giant tears—equal parts passion and liquor—were streaming down his cheeks. He was a very emotional screamer.

Everyone looked at each other, very confused. Someone in the crowd shouted at him, "I don't know what the hell you just said, but I second that motion." Before the Swede could respond, one of the paramedic-firefighters sitting in the front bolted out of his seat and screamed, "This is total bullshit! Our bitch mayor has treated us like a bunch of redheaded stepchildren, and I'm sick of it. All we want is parity with the cops. I do 10 times as much work over the course of my shift than any cop. We're so underpaid, I couldn't afford to take a vacation this year."

Our union's vice president was an older gentleman who sat through the medic's speech with a look of total ambivalence as he smoked a cigarette and drank vodka over ice from a highball. He allowed the medic to finish his speech, then turned to him and said, "Don't try to bullshit me, Jesus. Everyone in this hall knows you went on vacation this year. For the love of God, son, you're the richest Mexican I know."

The medic firefighter couldn't help himself and burst into laughter only to be interrupted by Colonel Frank shouting, "Point of order, Mr. Chairman, the floor wishes to be recognized." Our union president smiled and said, "Mr. Chairman recognizes Frank on the floor." Frank stood up. He had draped a horse blanket around his shoulders. He spun around several times, modeling and showing off his squaw shawl. He was sporting a shit-eating grin across his intoxicated face as he surveyed the group. "What am I offered for this beautiful piece of Native American tapestry? Let the bidding begin at $500." The crowd broke into laughter and began to pelt Frank with beer cans. For all practical purposes, my first union meeting had come to an end. I don't remember taking a vote on anything that night.

☠☠☠☠☠☠☠☠

Family history repeated itself a couple years later when my brother John joined the department. He had just turned 18. His only higher education came from a weeklong stint in a court-mandated driver education class. My brother had accumulated enough traffic violations to have his driver's license suspended. This is a true achievement when you consider he accomplished this feat before he was old enough to legally operate a motor vehicle. My brother has always been an overachiever and quite mature for his age. He gets his driven nature from our father.

Following the traditions of the fire-service cult, one of the first purchases my brother made was a sports car. A few years after he bought his red Trans Am, we were standing in my father's front yard discussing the next construction project he had planned for his two sons.

My dad asked John, "How fast does a car like that go?"

My brother gave the clinical response of "142 mph."

Dad processed this for several seconds before replying, "That sounds pretty specific."

John said, "Do you really want to know how I know?"

Dear old dad said, "I'm sure I won't hear anything more entertaining this week. Let me have it."

"One day I got sent to fill in at Station 19 (our department's airport station). That afternoon, I was washing my car when the station captain came out back. He was admiring my car and asked the same question you just asked: 'How fast does that thing go?' I told him I didn't have any idea because the speedometer tops out at 85 mph. He told me if I was interested, we could find out later that night. After that night's dinner, he picked up the phone to the control tower and asked if it wasn't too busy, could we borrow one of the runways for 10 minutes. The tower told him that would be fine. Before my captain hung up, he asked the tower to send over a traffic cop with a radar gun because we needed to test a new piece of aircraft firefighting apparatus. I got to run up to my car's top speed of 142 mph on Runway 26 Right."

My father stared off into the distance shaking his head. Before he actually spoke, I thought he was going to deliver a lecture about closing the airport so the firemen could do time trials. Instead he invited us into the backyard, where he produced two sledgehammers and 100 feet of footings he wanted removed. As he went into the house, I heard him giggle over the steady boom of iron into concrete.

☠☠☠☠☠☠☠☠

When I first began my career, the sky was the limit. Everything was brand new. It was like taking your first bite of a strawberry. We were heroes. The people loved us. Even the cops would sit idly by while we broke every traffic law ever written and axed our way into other people's property. Some days it felt like the city existed for our own amusement and fascination, and occasionally to help us pursue self-actualization through near-death experiences. Not many people have a job where they can watch their younger brother get blown out of a building in a spectacular fireball.

During the early 1980s, my brother and I were firefighters working on the same shift. I was assigned to Engine 23. John worked the next first-due area over at Station 59. During this era, I worked on a fire truck that was staffed with five firefighters but only had seating for four. My standard riding position was on the large rear tailboard. Like most hazardous activities, it was extreme fun getting to ride while standing up on the ass end of an out-of-control fire engine.

One summer evening, we got very busy due to a monsoon storm. Most of these calls were for downed power lines, tree fires and other weather-related mishaps caused by heavy rain, lightning strikes and driving winds. Scorching summer heat pounded the city by day. At night, moisture-bearing winds blew into town. The combination of moist ocean air and magma-like ground heat creates spectacular thunder and lightning.

We were babysitting electrical lines that had been blown off a power pole when we noticed smoke and fire several blocks away. The group of us jumped on the truck and headed for the rising smoke. The storm was directly overhead. As we rounded a corner, I could see over the top of the rig that a large palm tree had been struck by lightning. Palm trees are full of palm oil and typically wear a highly flammable beard of dead palm fronds. As our engineer spotted the rig in front of the burning tree and engaged the truck into pump gear, a large clap of thunder exploded. We were temporarily blinded and scared shitless by the intense flash of lightening that immediately followed. Regaining our wits, we stretched a red line off the hose reel and quickly knocked down the blazing tree.

Even in a driving rain, we used the better part of our 500-gallon water tank to extinguish the burning palm. We wound up the red line and headed up the street to fill our tank from the hydrant adjacent to a large vacant lot. We spotted the plug, made our hose connection and opened the hydrant. The group of us was standing next to the truck waiting for the hydrant to top off our tank when a powerful detonation of thunder physically knocked us to the ground. The air was charged with the smell of cooked earth, and in the distance, large, black clouds swelled with strings of white light. Rain drops the size of locusts pelted our turnout gear. For an instant, I thought maybe Jesus was coming for us.

After determining that none of the crew had been blown apart, we disconnected from the hydrant, mounted our 1976 American La France 1,500 GPM pumper and headed toward the relative sanctity of our station. On our trek home, we got popped for a structure fire a couple miles south. The southern sky was a mass of rolling black thunderheads cut by a 40-foot geyser of orange flame and churning brown smoke. As we raced toward the scene, I had the uncanny feeling we were responding toward a hell-bound choo-choo train. None of my friends who went to college were having this much fun.

Because I rode on the bumper, I couldn't hear any radio traffic concerning the fire. All I got was the roar of the engine and the constant whoosh of our 16-ton fire truck feathering the edge of a hydroplane. I could feel the torque shift as our engineer downshifted the transmission to slow down the truck. He swung the rig wide left, and then spun the wheel back to the right. This slid our truck through a right-hand turn that took us down the street where the fire awaited us. As we came to a stop, I spotted Engine 59 a few hundred feet ahead of us. They had stopped to lay a supply line. I unsnapped my safety strap, leaned over to open the SCBA compartment on the left side of the truck and pulled my mask from the compartment. I placed my arms between the chest straps, raised the air pack over my head and dropped the device onto my back. My partners had stood up from their jump seats, taken their masks from the transverse hosebed and executed the same procedure.

As I pushed my helmet and Nomex hood back to pull my face piece over my head, I looked up and saw Engine 59 laying a snake of 3½" supply line as the truck disappeared behind a cloud of diesel exhaust. After getting my face piece on, I opened the valve that activated the air flow, pulled my hood and helmet back into place, cinched the straps on my SCBA harness, put on my heavy leather firefighting gloves and waited to go. I could see the back of Engine 59's nozzleman as he put on his face piece. The group of us was cloaked in our fire-resistant robes, breathing air from pressurized bottles and waiting to take communion with the orange God of Heat.

The lightning, dancing flames and rain combined with the Engine 59's staccato preparations created a scene worthy of "Alice in Wonderland." The engineer bolted from the cab and ran toward the rear of the truck while the firefighter hopped off the side step and wove his arms through the loops of the preconnected 1½" attack line hanging from the transverse hosebed. In one smooth motion, he turned toward the conflagration and pulled 150 feet of hose out of the bed and began to advance it. His engineer was making pull-backs of 3½" supply line from the rear hosebed. He had elected not to apply the hose clamp to the still-dry supply line his plugman was hooking to the fire hydrant. In a few seconds, this line would be flowing 700 hundred gallons of water a minute. The engineer was racing to connect the end of the supply to the rig's intake before the plugman got him water. Engine 59's engineer would have to contend with 100 lbs. of water per second if he didn't beat the plugman with his connection. His second pullback produced a coupling that joined the 50-foot sections of hose. The engineer unscrewed the coupling and threw the female end of the hose into the hosebed. He pulled the line around to the side of his rig, where he screwed the male end of the supply line into the pump inlet.

As our vehicle accelerated toward the scene, we passed Engine 59's plugman as he began to twist open the hydrant. I looked down at the ground and saw Engine 59's supply line coming to life with water as it raced us to the scene. I could see their nozzleman had his attack line flaked out in the front yard. He threw his nozzle down and attempted to open the front door. It was locked, so he took a step backward, raised his right foot and drove it into the lock above the door handle. The jamb splintered, and the door gave a few inches. He reloaded his leg and blasted the door again with his foot. It flew open just as his attack line tensed with 200 lbs. of water pressure, causing it to roller-coaster across the front yard.

As we pulled up to the scene, we could see two houses well involved in fire. The roof of the house Engine 59 was preparing to attack was totally engulfed in fire. Heavy flame pushed out its gabled ends and from the eaves of the roof along the perimeter of the house. The flames curled up and over the eaves, converging across the shingled roof like a wave. It then formed a 20-foot-wide column that shot straight up into the sky. It looked like the hair on one of my sister's troll dolls.

As Engine 59's nozzleman prepared to enter the rapidly vaporizing house, thick black smoke pulsed from the top of the front door. Right before he disappeared inside, the nozzleman turned his head in my direction. I didn't need to see his face to know it was my brother.

This entire scene took place in the same amount of time it takes the average person to sneeze twice. It is amazing the amount of information the human brain can process in a short period of time. We had been assigned to attack the burning house next door. I pulled a second 1½" attack line off Engine 59. I was waiting for water in the front yard of the house we'd been ordered to attack, watching fire and smoke push out of the windows in the front of the house. The fire was generating too much heat for us to make entry. We were going to have to knock down a lot of fire before we could get inside. From behind me, someone opened a line and was focusing a stream of water through the burning windows on the other end of the house. I turned and saw my partners from Engine 23 as they began working on the large volume of fire threatening the next house over. Two houses were well involved in fire and the houses on either side of them were starting to smoke. You could hear the fire laughing at the rain.

Engine 59's engineer had just charged my line. Before I started flowing water, I turned to take a final look at the house with the roof that sported a troll-doll hairdo. The free burning column coming off the roof was no longer going straight up in the air. It had taken a corkscrew shape and now resembled a burning tornado. The smoke was curling out of the top of the front door and down both sides of the doorway. The center section of the doorway remained clear. The smoke had changed from black to a dirty yellow dissolving into a light coppery brown. I still remember thinking to myself, "I don't know what my brother's doing inside there, but he sure as hell isn't putting any of that fire out." Before I focused my attention on my own little piece of our burning planet, I saw the smoke that had been puffing out of the doorway had reversed its course and was retreating into the burning house. This is an omen of bad things to come.

After two or three cycles of smoke pushing out and then retreating back into the open door, I heard a muffled explosion from within the house. The smoke-scorched windows exploded, and a rolling ball of smoke and flame shot into the front yard. It quickly disintegrated, giving birth to a firefighter. My brother was tumbling through the air, a blur of knees and elbows in his bright yellow turnout jacket. The backdraft had blown him out of the house and halfway into the front yard. I stood in apt amazement, obsessed with the odds that I had almost been struck by lightning (twice) a few minutes ago, and now my brother had been blown out of a building. Maybe Jesus was coming for the Brunacini brothers.

After John came to a stop, he got up off the ground, readjusted his SCBA harness and smoldering fire helmet and casually walked back to the house that had just so harshly evicted him. When he got to the front door, he bent over, grabbed his attack line and pulled hose out until he came to the nozzle at the end of the line. Then he opened the nozzle and began knocking down the fire blowtorching from the front door. I turned back to see my partners on Engine 23, who were also transfixed by the event. We looked at one another, shrugged our shoulders and went back to work. Our careers were as good as a Johnny Cash song.

Chapter 3
Girls Gone Wild

The American fire service is one of the most visible organizations on the planet. We paint all of our equipment in loud primary colors. To ensure we don't go unnoticed, we adorn our apparatus with an array of lights that flash, blink, oscillate, rotate and strobe. We accessorize our rolling beasts with sirens and air horns that scream, "Notice me!" It's a wonder more gays aren't drawn to a career built around such flamboyant rolling stock.

We build our fire stations in the center of every neighborhood in the community. We deliver service 24/7 uptown, downtown, in the suburbs, on the wrong side of the tracks or anywhere else someone punches 911 into a phone and asks for our help. We are everywhere.

In a way, the fire service is responsible for rocketing the ancient concept of the clan to cult level. We wear uniforms adorned with patches, brass buttons, gold ropes and obscure trumpets that signify our level within the cult. The senior members of the group have been known to crown this ensemble with a chrome-plated, penis-shaped (just the tip) helmet. Our symbol is the Maltese cross, and all of our ceremonies begin and end with skirt-wearing men playing bagpipes. Most mental-health professionals would define this level of pageantry as a tad bit cultish.

The true test of any cult lies in how they recruit and assimilate new members. Over the centuries, we have refined and standardized the process we use to fill our ranks. In our early days, we sought out young, hyperactive males with strong bones and teeth. We would then process them with a trial-by-fire initiation. These rituals began with dressing the new recruits in the protective robes of the society. After introductions, a nearby building was set ablaze, and the new members were locked inside. This process was repeated until the elders had eliminated any of the pledges who screamed like sissies, ran out of the inferno, cried for their mamas or Jesus, or asked too many questions. This left the cult with fledglings who were fire-tempered on the outside, but still soft and malleable on the inside.

The next gauntlet the newbies faced was the socialization process. The youngsters joined the general workforce where they would march a few steps behind and to the side of the senior firefighting wizards and warlocks. Introduction to station life revolved around the young learning the finer points of station cleaning, ladder sanding, phone answering and dishwashing. A few hours of every shift were reserved for hazing. This included (but was not limited too) stripping the newest member naked, tying him facedown to a backboard, sticking a jalapeño pepper into his barely mature anus, placing him on the truck's hosebed and driving around town. If this isn't cult-like behavior, I don't know what is.

Imagine the shock and terror that befell our fire-service cult when a higher-ranking cult (the Supreme Court to be exact) ordered us to begin hiring women. We were a closed society of manly men. Like the Roman legions, cowboys of the wide-open range and the Greek navy that came before us, we were intrepid and resourceful—and we could tie knots. But more importantly, we were all male. Our only fear was to be excluded from our cult. Now the courts had mandated that women were perfectly capable of performing our craft. The horror!

Being the chief's kid was small change compared to the trials and tribulations of being a firefighter with a vagina. I am a man working in a man's job. Some guys can be pretty demeaning about the whole gender issue, and a lot of the guys who are nice to the girls are only doing so in an effort to see their titties.

The hot-button issue regarding women in the fire service was whether or not they could physically do the job. Enlightened fire-service leaders handled this concern by designing a physical-agility test that is truly job related. During the course of my career, I've worked with a few female firefighters who were stronger than many male firefighters. The problem for the freakishly strong girls is finding on-duty workout partners. No self-respecting man who fancies himself a Viking warrior wants to strip plates off the bar when he follows a girl on the bench press.

Many of the older members had quite a bit of trouble accepting females performing non-secretarial functions. It was one thing to have to go inside burning buildings with them, but quite another to spend the night in the station dorm with them. This was also an insurmountable issue with many of their wives.

There was widespread fear that women would shut down our he-man firefighter fraternity/club/cult. Gone forever would be the days we could blurt out vulgar words, openly view German porn in the dayroom and boast about the year we spent driving Crystal Gale's tour bus by day and "knocking the bottom out of her pussy" by night. (A fellow firefighter swore to me he fathered a child with 1980's country music icon). We were so convinced women would wreck the social fabric of our cult that we hated them before they came through the door. Like most absolute beliefs any cult holds sacred, we were wrong on this one, too.

I worked with an old Marine named Earl for a little while when I was a young firefighter. Later on in my career, after I first made captain, I roved into Earl's station for a couple of shifts when his regular captain was on vacation. I was in the captain's office talking to the ladder captain, who also happened to be a fill-in guy. His name was Butch, and he was a very nice man. We were shooting the breeze when Earl came in to find out which one of us would be his captain. Earl kept automatic weapons in his locker. He had a severe crew cut that helped define him and what he was all about; the U.S. Marine Corps tattoos on his forearms were merely exclamation points.

"Hi Bruno, how you doing today?"

"I'm fine, Earl. How about you, my brother?"

"I just came in to find out which one of you is going to be my captain while Freddie is on vacation."

Butch asked Earl which rig he worked on, the engine or ladder. Earl looked at Butch like he had just detected a bad smell.

"Who are you?"

"My name is Butch."

"You ever kill a man?"

Butch was surprised by Earl's line of questioning. "No, I haven't."

"I didn't think so. Bruno, I can't work for this no-neck. Please tell me you're my captain for the next couple weeks."

"Earl, I've never killed anyone either. What difference does it make?"

"I think you could if you had to, plus you're diseased. You make me laugh, and as we all know laughter is the best medicine."

My dad met with all the troops at least once a year to shoot the shit. I was working with Earl when it was our turn to meet with Big Al. At the end of the meeting, the fire chief asked if anyone had any questions. Earl raised his hand and waited for his Supreme Imminence to call on him. Earl was very polite that way. "Chief, I have a question about qualifications for the engineer's position."

Earl wasn't always easy to read. Big Al looked pensively at Earl and asked what his question was.

"Our department has been hiring women for about seven years now. Some of them are starting to make noise about taking the engineer's test. Are you going to allow that?"

The chief couldn't help himself. "Earl, why wouldn't we let someone take the test if they're eligible?"

Earl was as serious as a torn perineum (that's "taint" for all you B-shifters reading along) when he said, "Chief, I mean no disrespect to anyone when I say this. It is both a medical and scientific fact that women have menstrual periods. The week before they start to bleed from their vaginas, they become very mean. Some of them can even be described as homicidal. I do not think we should allow someone who is not in their right mind to operate a 16-ton piece of equipment while they drive at speeds in excess of 55 mph and have the legal authority to break most traffic laws. They could injure or kill the public or our own members. You're the only one who can keep it from happening."

I do not remember what my dad's reply was. It certainly couldn't have provided Earl with a satisfactory answer to his question.

☠☠☠☠☠☠☠☠☠

Within the cozy confines of my own department, none of the female firefighters has ever taken a grievance outside of our organization. There is a long line of people who would gladly (and inappropriately) take credit for this anomaly, but the credit fully lies with the female members of my department. The first women to join our department were very patient. This group of early pioneers simply wanted to be firefighters. They knew it wasn't going to be easy breaking into an organization that had been exclusively male for its entire history. After a while, we all got used to one another and figured out female firefighters were a lot like their male counterparts. The girls share the same spectrum of human traits — smart to dim, pretty to not so pretty, strong to weak, sane to nuts, team players to shit-stirrers, and fun to serious. If we exclude sexual genitalia, the women are exactly like the men.

With very few exceptions, female firefighters want to get along, do their jobs and be considered equal members of the cult. The female firefighters whom I've worked with didn't join up to create a commotion or take the department to court in search of a big payday. When faced with some type of problem, they've processed it through the normal chain of command, as the following incident demonstrates.

Suzie, a female paramedic-firefighter, roved into a multi-company fire station that housed 10 firefighters and three pieces of equipment. The individual sleeping quarters in the station dorm were segregated with lockers on either side. Each "rack" contained a twin bed, a chair and a television set. The bed was positioned against the backside of the locker for the "rack" next door. The entrance to each sleeping chamber was covered with a heavy cotton curtain. This gave each firefighter a modicum of privacy. It was also a big improvement compared to the way the sleeping and bathroom facilities were arranged when I began my career.

Prior to the inclusion of women in our service, most fire stations could best be described as sparse and utilitarian. Dorms were one common room crammed to capacity with twin beds. They were more cow pen than sleeping quarters. It was common to have a dozen firefighters snoring, farting and having Vietnam flashbacks in a dorm smaller than a two-car garage. I began my career in 1980 and found it odd that the Phoenix Fire Department had yet to invest in a bathroom stall. I still have post-traumatic stress from what I witnessed when I walked into my first station bathroom. The three toilets stood naked against the wall facing the door. The engineer was cleaning what looked like last night's exploded green chili burro (enchilada style) off his ass with a handful of wadded-up blue windshield towels. In retrospect, I owe this man a great debt. My first decapitation paled in comparison to the vile image of his foul ass. Back to Suzy...

After a busy shift of racquet ball, extinguishing car fires and treating ill diabetics, Suzy crawled into bed hoping to get some well-deserved sleep. As Suzy lay there willing sleep to come, she took notice of noise from a nearby bunk. Suzy listened for several minutes before determining that one of her fellow firefighters was in the throws of self-inflicted passion. She quietly giggled to herself and did a mental rundown of who the bold masturbator might be. It didn't take her long to figure out that Mr. Onanism was the firefighter on the other engine company. Suzy found it odd that a man-boy pretty enough to play a supporting role in a big budget gladiator movie would be jerking off in the cult clubhouse. Suzy rolled over, put her pillow over her head and attempted to delve into sleep. But sleep didn't come for some time because Spartacus spent 45 minutes grunting and moaning while his mattress made squeaky noises that synchronized with his jerking motions.

The next morning, Suzy woke up, got dressed, brushed her teeth and loaded her gear into her car after the oncoming shift relieved her. She didn't bitch, whine or file a complaint with the federal courts. The next shift, Suzy roved back to the same station. After she checked out the rig, stowed her gear and made her bed, she sought out her captain for a private conversation. Suzy described her prior shift's 45-minute self-love soundtrack. Her captain reacted as if his personal physician just told him he had untreatable parasites eating his brain. Sensing her captain had visions of her bringing on a full-blown inquisition, Suzy clarified her position.

"Listen Skip, I really like working for you. I like the station and the guys. I don't want to make a big deal out of this. I just want to be able to go to bed at night without any jacking-off noises. I'd be happy if one of two things happens. My main preference is for Spartacus not to jerk off while I'm around, but if it's something he absolutely must do, I'd settle for him finishing up his business a lot quicker."

☠☠☠☠☠☠☠☠

My own department/cult hired our first woman in the late 1970s. She ended up washing out of the training academy. It wasn't until a few years later that women had navigated their way through the testing process, successfully completed recruit school and became sworn members of the workforce. I remember going off work one shift during this era. It was around 0730 and the C-shifters were coming on duty. I was in the apparatus bay of the station talking to one of my fellow B-shifters. This cult member was a veteran firefighter who had been in his share of burning buildings. His pastimes included power lifting and finishing concrete. He always spoke in the elevated tones of an outside voice because he was half-deaf from riding in the hip pocket of screeching diesel engines and ear-curdling sirens. The two of us were talking in the apparatus bay when a young girl walked in our direction from the parking lot behind the station. She looked like a very attractive 16-year-old gymnast. She stood 5'4" and couldn't have weighed more than 100 lbs. She was wearing a white athletic shirt and a pair of those blue Dolphin jogging shorts that were the rage in the late '70s. Her arms were loaded with a turnout jacket, bunker pants and a fire helmet. As she walked up to us, my partner said, "That's so sweet. You're helping your father with his gear. Which one of the C-shifters is your daddy?"

The sweet, innocent young girl dropped the gear out of both arms, screwed her face into a wicked witch scowl and shook her finger in my buddy's face as she seethed through clenched teeth, "Listen asshole. I'm a firefighter assigned to this station today. I'm not some C-shifter's little girl. I showed up today to work, not to take any of your lame shit."

When she was done with him, her gaze shifted to me. "Good morning, Nick," she said. "And a good morning to you, Darla," I replied. After she stormed off, my buddy gave me a puzzled look and said, "Jesus Christ, Bruno, I really thought she was bringing her dad's gear in for him. She's so tiny. I bet she shits little bunny pellets." It was a brave new era.

I had first worked with Darla several years earlier when I was assigned to a firehouse sandwiched between the city dump and dozens of trailer parks. The sound of banjo music could be heard throughout the first-due area. I was the senior firefighter on B Shift. Our station was a training station for new firefighters. Every three months, a new probationary firefighter would cycle through on each of the three shifts. Darla was a brand new C-shifter fresh out of the training academy.

One morning I was getting off work and had gone out to the apparatus bay to take my gear off the rig when I ran into Darla. She had been assigned to the station for a couple of shifts, but short of a brief introduction we hadn't had a chance to have an actual conversation. I approached Darla, and I could tell she was uncertain how this encounter with a large and unkempt B-shifter would go. It's a pretty good bet that during the infancy of her career, she'd suffered her share of idiocy from some of our more insecure male cult members.

I broke the ice by opening the ALS compartment on the rig and asked her, "Darla, have the C Shift medics shown you how the daily check of the ALS gear is performed?"

Seeing I was actually trying to help her out, her face softened. "No," she said. "They told me to check just the BLS box, and they would take care of the ALS stuff."

"Excuse me for saying so, but they are a couple of thin-wristed morons. You should know how every piece of equipment on the rig works. I'd be happy to show you." Darla beamed. "I'd really appreciate that, Nick."

"Good," I replied. "Let's start with the nitrous kit."

I pulled the nitrous oxide kit off the truck. It was a device that delivered a mix of oxygen and nitrous oxide (laughing gas) to patients in extreme pain. Darla followed me as I carried the box into the storeroom just off the apparatus bay and set it down on a large workbench. I opened the box and showed her how to hook up the cylinders of nitrous and oxygen. I gave her a quick overview of the unit's basic function, the proper operating pressures, how to adjust the flow of each gas, and how to change out the bottles when they got low. During my overview, I took the device apart. When I finished showing her how everything worked, I asked if she wanted to put it all back together. Within a couple of minutes, she had the kit reassembled.

After she finished, I reached over and turned on the valves that controlled the gas flow and told her, "Now we have to make sure there are no leaks. These units are notorious for leaking." I grabbed the face mask, took a large hit of gas and passed the mask to her as I held my breath. She looked at me like I had just involved her in a bank robbery. I exhaled and said, "You really need to test the quality and temperature of the mix. EMS gave us standing orders to test these units every morning."

She considered this for a few seconds, grabbed the mask and took a long pull of gas. After the third hit of gas, we were both sitting on the floor of the storeroom giggling like little girls. After the 10th or 12th go round, she grabbed my hand as she went for the mask. She pulled my hand, with the mask in it, to her face and started really huffing.

I felt like I was levitating several feet off the floor when I noticed the fine pelt of blonde hair on Darla's arms. I became transfixed for what seemed like a really long time by the hippie-girl braided leather bracelet she wore on her left wrist. My eyes followed the cord of veins that ran along her small and muscular forearm, up to her bicep and toward her shoulder until my gaze landed on her pert breasts. I looked up to find she had caught me staring at her boobs. This brought on a fresh laughing jag. Because she had been Bogarting all of the gas, my head had begun to clear, and my thoughts turned south. Before I could take advantage of the situation, Darla's captain barged in on us and asked what we were doing. She was sprawled out on the floor laughing like an imbecile who just watched a monkey perform a magic trick.

Prolonged use of the nitrous caused the bottles to form a thick coat of frost where they connected to the main unit. Darla's captain looked at the 1-inch layer of ice that had built up on the bottles and shook his head. He sent a still-giggling Darla into the station. When she was safely out of earshot he told me, "I have a pretty good idea where your little morning check was heading. I would appreciate it if you would refrain from getting our probationary firefighter so gassed that she doesn't make very wise decisions."

I feigned great offense and asked, "What is that supposed to mean?"

"What it means, Mr. Morally Bankrupt B-shifter, is this: The next time you want to test the nitrous kit, do it with your own probie."

Darla finished her time at our station without any further career guidance from me. A few months later, the training academy sent us our first female probie. Her name was Vicky. She was a pleasant person and a capable firefighter, but very competitive. This is a normal characteristic possessed by the people hired into our organization/cult, but Vicky took it to an obnoxious extreme. She turned everything into a contest. Working with her became so tedious that I withheld demonstrating how the nitrous kit worked.

Vicky was a real pain in the ass. Maybe she rubbed everyone the wrong way because she felt a great deal of pressure to excel in a traditionally male occupation. At the end of the day it really didn't matter—no one likes working with a person who spends most of their talking time harping about how good they are. Every human activity became a race. Vicky was the first one to the rig when we had a call. Vicky could take a patient's full set of vitals faster than anyone else on the truck. Vicky could pull an attack line as fast as the rest of us. Vicky, Vicky, Vicky… Vicky hadn't figured out she was a member of a four-person crew. Vicky also hadn't been in a real fire during her young career.

It is a universally held truth in our cult that nothing exceeds the joy, terror, fun and general rush of holding a nozzle flowing big water onto an out-of-control blaze inside a burning building. My department staffs our fire engines with four people. Back then, our standard response to structure fires was two engine companies and one ladder. The vast majority of structure fires are extinguished with one attack line. Twelve firefighters arrive on scene, and most of the time only one of us will have the lion's share of self-actualization. This places a premium on getting the nozzle (or as our ancient Greek brothers called it, the nob). There are 300-year-old accounts of fire-fighters fist-fighting in front of burning buildings for the right to put first water on the fire. As stupid as this was three centuries ago, it continues to happen today.

Over time, a firefighter learns the proper sequence to follow when competing for the coveted nozzle. Just because a firefighter pulls the initial attack line doesn't ensure they will end up with the nozzle. A lot can happen between pulling the line off the rig and actually flowing water. Many new members make the mistake of not putting on their SCBA face piece before they pull the line off the truck. This means they have to stop at the front door, drop the nozzle and go through a 10- to 20-second process of donning their mask. This procedure requires loosening your helmet strap, pulling your Nomex hood back, putting your chin in the bottom of your mask and pulling the webbing over your head, turning on your air supply, adjusting the mask to make sure it isn't leaking, reapplying the hood, putting your helmet back on, and tightening its chin strap. This is a much easier to do with bare hands as opposed to bulky firefighting gloves. The problem is if you do it bare handed, it just adds another piece of safety gear you have to screw around with before you tuck the charged line under your arm and take the nozzle into your hands, claiming it once and for all as yours.

Non-cult members might be asking themselves, "What's the big rush? You're preparing to go inside a burning building. Who in their right mind wouldn't want to take a couple extra seconds to ensure their safety gear is properly configured?" This is an excellent point. We routinely operate in places that will kiss any piece of exposed skin with a second-degree burn. I have also seen more than my share of firefighters who rendered themselves functionally blind because they were in such a rush to put on their headgear that they pulled their Nomex hood completely over the lens of their SCBA face mask. Firefighters will search out and fight over a nozzle much like bulls sniff out and fight over cows in heat. Bulls do it because their biology programs them to; firefighters exhibit these behaviors because at our very core, we are self-destructive adolescents.

One fateful day, Vicky had an experience that caused her to reconsider the grand vision of her place in the universe. Our company was dispatched on a house fire. While we were responding, Vicky got dressed 13 seconds faster than I did. Our engine company was the first unit to arrive on the scene. As luck would have it, there was a fire hydrant right in front of the burning house. Vicky came off the truck, threw on her SCBA and pulled an attack line in the record-setting time of 27 seconds. She had the line flaked out in the front yard, waiting for our engineer to give us water. I threw my bottle and put my face piece on as I walked across the front yard to hook up with her. As I joined Vicky, our attack line was charged with water. She was standing there without her face piece on, holding the fully charged line as she waited in front of the burning house. She had spent the last couple of weeks beating all of us. Now when there was real work to do, the best she could muster was that of a second-place apprentice.

She pinched the line under her arm and started the task of donning her mask. I don't know if she considered me chivalrous or simply stupid as I patiently stood by while she masked up. It was painful to watch. She kept the charged line pinched under her arm while executing a maneuver that required the full range of two hands connected to both arms. It looked like she was born with one normal arm and a flipper. I could see a look of triumph in Vicky's eyes when she was finally geared up and had managed to keep the line. She advanced the line to the open front door with me behind her. Eight minutes later, we would exit the structure through the same door, and Vicky would have a brand new look in her eye.

The fire started in the back of the house when the tenants decided to use gasoline as a floor stripper. Both members of the cleaning crew were fortunate to receive only minor burns. The house wasn't so lucky. Just a gallon or two of gasoline will cause the fire to burn much hotter and faster. As Vicky and I made our way into the house, we could hear and feel the fire roaring behind the interior walls. The fire was pushing black smoke across the living room ceiling and out the door we had just entered. It looked like the ghosts of a thousand pissed-off bowling balls rolling across the ceiling. Just a few steps inside the house, it was too hot to stand. As I knelt down, I could see Vicky was still upright. I couldn't see her from the shoulders up because her head was lost in the smoke. I reached up and pulled her down. At that moment, I realized she hadn't opened the line. Several seconds later, the smoke above our heads flashed over in shades of red and dark orange. I reached around her, opened the line and played it into the burning ceiling. A minute or so later, the hundred gallons of water we had put into the living room ceiling space had knocked down the fire in our immediate area.

38

I got to my feet and grabbed Vicky. She had both arms wrapped in a death grip around the line. I picked her up and carried her through the kitchen/living room area toward the rear of the house where the main body of the fire awaited us. We were joined like a pair of figure skaters. Vicky was getting twitchy. She thrashed like a cat that didn't want to take a bath. It was still warm in the living room, and Vicky made a lovely heat shield.

Each fire has its own personality. Most structure fires are hot and smoky with little to no visibility. You generally don't see much flame. If the immediate fire area is vertically ventilated before you actually find and extinguish the blaze, the smoke and heat rise up and away. This makes for a very beautiful fire. Sometimes you can see all the solid fuel vaporize into gas. Sofas, chairs, wallpaper, children's toys and everything else in the fire area retain their basic shapes, but their surfaces radiate an aura of transparent energy finished with a blush of flame. Nature is one serious bitch.

Vicky and I were about to round a corner when we found another kind of fire—a blast furnace. Blast-furnace fires are mean and angry. We made the corner of the living room and came face to face with a flame front that filled every square inch of the back half of the house. My natural reaction was to get my little Tinkerbell and me as low as possible and open the nozzle.

Stuffing 125 gallons a minute into the fire initially didn't have much of an effect. We could smell the gasoline cooking off through our masks as we whipped the nozzle in large circular motions. The straight stream of water completely disappeared into flame. As the fire took our water, it began to fade. Eventually the water darkened down the fire, replacing the roaring flame with steam, dirty smoke and lots of carbon monoxide. I stood up, taking Vicky and the line with me. Venturing deeper into the general fire area, we could hear the fire crackling somewhere ahead of us. We couldn't see shit. I was moving Vicky by pushing her and occasionally carrying her in the direction of the still-burning blaze.

The on-the-job mentoring I was providing Vicky wasn't a whole lot different than the education my cult elders provided me. On more than one occasion, a guiding hand hit me in the back of my helmet with a flashlight or ripped a nozzle or tool from my hands because I wasn't performing a task to the correct specifications. This type of teaching goes with the territory. It isn't possible to have a traditional question-and-answer period when you consider that the inside of a burning building makes for a pretty shitty classroom and our job routinely kills more than 100 of us each year.

We stumbled and tripped over a steady stream of smoldering contents until we reached a wall. We followed the wall until we ran into a corner. This led us into some kind of closet or bathroom. It was difficult to tell while dragging Vicky and the line. The smoke was still heavy, and I couldn't see much past a few feet. To make matters worse, the steam and our heavy breathing had fogged up the insides of our masks, making the world around us look as if it were drenched in dirty syrup. The two of us were stuck in a maze that was filling with fresh black smoke and heat. We were retracing our path back toward the front of the house when I bumped into a wall. My shoulder knocked a hole in the fire-weakened partition. The hole quickly filled with fire. Vicky, a couple steps ahead of me, was dragging the line back toward the living room. I grabbed her from behind, put my mask to her ear and said, "The fire is burning behind that wall."

Communication in a burning structure is difficult at best. Imagine wearing a breathing device that you must literally suck air from when under heavy exertion. Your ears are covered with three layers of Nomex. Verbal communications sound like an asthmatic hostage pleading with his captors through a pillowcase that has been duct taped around his head.

Vicky couldn't understand my screaming through the mask, and she wanted out of the room. I was going to have to pick her up again. I put my arms around her and opened the nozzle she was clutching. Once I got the two of us pointed in the right direction, I began knocking large holes in the wall with my foot and Vicky. She worked the nozzle as if her life depended on it. A few minutes later, the fire was out. Vicky walked out of the smoldering house under her own power.

Vicky and I stood in the front yard. Through the burned-out shell of the house, we watched as the second-due engine company mopped up the fire in the patio area and backyard. We shed our SCBAs, helmets, hoods and turnout jackets. Vicky was having some type of internal dialogue. Several times she looked as if she was going to speak but stopped herself. After a while, she just stared at a spot on some distant planet. This is a normal reaction when you are finally jumped into your new gang. Maybe part of Vicky's front-yard confusion stemmed from the notion that she really wasn't a member of the cult until just now. Maybe she figured out that the first four months of her career were insignificant when compared to the last 8 minutes. Maybe she was just in a daze. Whatever the reason, Vicky became much easier to work with.

In 1981, women climbed over the wall and became a regular part of my cult. After a couple of years, the novelty of women on the job wore off, and life went on in a coed fashion. The mixing of the sexes would take a more personal note when Candi, my baby sister, decided a career in the fire service was for her.

Candi tested twice before she got the job. During her first attempt, she ran out of time before completing the physical-agility test. Sis spent the previous five years working with her first husband on a feedlot. She was farm-chores strong. After her divorce, she came back to town and decided to take the firefighter's test. My brother John and I had been on the department for more than a decade. John sat Candi down and asked her if she was aware she was choosing a career that was more difficult for females. After it became clear to him that our sister was serious about this, he told her, "Nick and I have earned reputations as competent firefighters, and dad is becoming the new messiah of the fire service. If you embarrass our family, I will punch your heart out." I don't think my brother's little pep talk impressed our sister.

Six months prior to Candi's physical-agility test, John asked her how she was preparing for it. Candi looked at him and rolled her eyes. Her expression made it obvious that she thought he was being a moron. She was 5'7", weighed 175 lbs. and could bench-press 250 lbs. She was used to doing man labor. She looked at him and said, "I am being personally trained by Janou Helson. I work out using a split routine. I do a push-pull anaerobic workout four days a week coupled with cardiovascular workouts the remaining three days."

Janou Helson had been a firefighter in a neighboring city and was a competitive weightlifter before becoming a firefighter. A year or so later, he went back to school, got a degree in physical fitness and opened a physical rehab center. This started taking up most of his time. It also filled up his bank account, so he quit his firefighting job to devote all his energy into his business. Candi showed John all of the scientific charts, diets and workout routines she was following. Now it was John's turn to roll his eyes.

"This is bullshit. The physical-agility test is a set of events. You should center your training around the test you will be taking."

Candi told John he didn't know what he was talking about; Janou was bigger and stronger than John and knew more about training than John ever would. The conversation ended when John told Candi, "If you don't pass this test and you embarrass our family, I own you."

The big day arrived and Candi went off to take her test. She had always done well with the academic side of things. She aced her written exam and only needed to pass the physical to earn an oral interview. All of Candi's friends showed up to cheer her on. I stood off to the side and out of the way with my brother. Just before she started, John looked at me and said, "Don't get me wrong, she's my sister and I love her, but if she doesn't pass this test she will wish she had never been born." Poor sis ran out of time less than 20 feet from the finish line.

The only reason my brother gave our sister a pass on failing the physical-agility test was that she gave it her all. Immediately after the test, sis went into renal failure and had to be hospitalized overnight. John gave Candi a full week to recover, then he sat down with her and asked what she planned to do. She reaffirmed that she still planned on becoming a firefighter. John reminded her of their earlier conversation. He understood how someone could easily be led to believe that if they did a push-pull cardio anaerobic exercise program, complete with special diets and charts and graphs, they could easily pass the physical agility test.

"You will not be doing any more of that stupid bullshit," he told her. "The next test is one year from now. Tomorrow I want you to start running for 30 minutes a day. Do that 5 days a week. Six months from now we will start training for the physical exam."

John was true to his word. One day he called me and asked if I would meet him at our parent's house after work. I met him the next morning and helped him set up a physical-agility course on the piece of vacant land across the street from my parent's house. We built a stand that held a 45-foot aluminum extension ladder. We flaked out and connected six 50-foot sections of 2½" hose. My brother set up a tree stump that was to be pounded on with a 12-lb. sledgehammer. John devised a contraption consisting of seven old truck tires chained together and tethered to a 10-foot loop of 2½" hose. The course looked like the set of a Mel Gibson movie.

My sister had just finished taking her second written entrance exam when I went over and saw firsthand what my brother was doing to get her ready for the physical. It was obvious they had been running through the course for some time before I arrived. She was beating on the stump with the big sledgehammer. John was yelling, "From the shoulder, only raise the hammer to your shoulder! Short, hard strokes! You have to hit it hard enough to ring the bell! Faster! Hit it faster! Attack! Girls don't know how to attack. You must learn. Attack it!"

Candi looked up at him and started to say something but John cut her off. "Shut the fuck up and just hit the stump until I say stop." John counted off 75 blows and told her to move on to the ladder raise. Candi let the sledgehammer drop out of her hands and started to walk toward the ladder. She was wearing a backpack my brother had loaded with four 25-lb. plates. She clanged when she walked. She reached the ladder and put her shoulder and hip into it as she reached up and grabbed the hemp rope used to extend the ladder's three fly sections. She was raising over 200 lbs. of ladders by pulling on the rope. She hesitated to catch her breath. My brother patiently stared at her. Sensing he was about to speak, she straightened herself upright and began to extend the ladder skyward. John had her do this three times before she moved onto the hose drag. She walked over to the 300 feet of 2½" hose and picked up the large nozzle attached to the end of it. She threw the nozzle over her shoulder, bent over and began to run. The farther she went, the more hose she dragged. When she got to the end, she was walking, but she had extended all 300 feet of the hose. She walked back over to us. She was saturated with sweat. John looked at her and said, "Now it's time to drag the tires."

She pleaded with him, "No, not today. I don't want to do the tires today." John replied, "I don't give a shit what you don't want to do today. You are done embarrassing our family. Pull the goddamn tires before I kick your ass."

Candi went over and grabbed the piece of hose attached to the tires and began to run. She had made it about 50 feet when she half fell and half-collapsed. John ran over to her. I mistakenly thought he was going to help her. Instead he started screaming at her. "Get up now! You are not finished. Get up off your ass right this minute! It's all right to have a pussy but you have got to quit acting like one. Now drag the goddamn tires the rest of the way!" John was irate. He was picking up dirt clods the size of melons and throwing them at our darling sister. She was covered with a fine silt of dust that was turning to mud. Her possessed brother had gone fully demonic. He had grabbed her hands and was pulling her, screaming at the top of his lungs for her to finish the course. Later on, I found out they did this little ritual three times a week for the six months leading up to the physical.

My sister passed her physical-agility test in half the allotted time. Several months later, she had her oral interview and was hired. She hasn't worked out with my brother since.

After Candi joined the fire department, she was assigned to a single-engine company house. The captain at this station liked my father, but detested my brother and me. He made this perfectly clear to little sis. He was a rawboned sodbuster type who fantasized about a return to the good old days. I figured it was just the early signs of Alzheimer's disease setting in. On my sister's first shift, he sat her down and told her that neither he nor his crew had ever worked with a woman before. He was worried about getting in trouble if he or one of the other guys offended her in some way. This was going on all over the American fire service. Women were suing departments over sexual-harassment issues. Many of these lawsuits also contained some type of criminal assault charges. Most of these cases deserved to end up in court. Even so, it was a good idea that her captain established a dialogue with her about station life.

"Candi, what would you say constitutes sexual harassment? We want to make your time here as pleasant as possible, but none of us really wants to undergo a massive personality change." My sister looked her 57-year-old captain in the eye and answered, "Penetration." Everyone got along just fine during her tour.

Chapter 4
The Piano Man

I do not remember what time it was, only that I was very comfortable beneath several layers of blankets when the lights came on. It was December, and it was a few degrees below freezing just outside the windows. The first thing I realized, after the lights and obnoxious tone brutalized me awake, was the air conditioner was on. In fact, both the a/c unit and swamp cooler were running full blast—a combination that produced the equivalent of lake-effect snow. The people around me slowly started getting out of bed. But at that moment, I did not want to get out of mine. My nose was numb. I had been pulled harshly from the sweet dream world of talking ponies and chocolate-and-vanilla swirl concubines when I realized I was at work. This realization filled me with a deep sense of loss that quickly turned to anger. I did not want to venture past the warmth and comfort of my cozy bed in Station 55's dorm.

We had been experiencing a long trend of 12 or more calls—and two to three hours of sleep—per shift; half of the calls occurred after 10 p.m. When you began to develop flu-like symptoms, you knew your shift was over and it was time to go home. The real downside to all of this was the type of calls denying us our sleep. A glut of late-night medical runs was causing our entire crew, with the exception of our booter (who was too new to know any better), to lose our last kernel of sanity. (Don't ask me where the term "booter" comes from because I don't think anyone living knows. It's common for our newest members to be addressed as "booter," "probie," "new guy," "pard" or simply "hey you" during the first year of their career.)

Eighty percent of our calls were medicals. These weren't your made-for-TV "first when seconds count" EMS calls. The vast majority of them were for crazy people, our homeless neighbors or people who had been sick since Jerry Lee Lewis married his 14-year-old cousin. Don't get me wrong: We routinely responded to folks who didn't have much time to make peace with their maker. Our first-due area had a prolific knife-and-gun club. We had our share of people who were shot, stabbed or bludgeoned to death (or near death) as a result of bad relationships. We also responded to a high number of car-vs.-semi "Red Asphalt" motor-vehicle accidents. Still, the majority of our EMS calls were substance abusers, chronic complainers and acute whiners. We had a cache of regulars who accounted for 5–10 percent of our medical calls.

Doris Mylar was the poster child for our regulars. Doris lived a couple blocks from the station in a deteriorating wood-frame hovel decorated with hundreds of framed jigsaw puzzles, many of them missing several pieces. She resided in this cozy little fixer-upper with her husband and mildly retarded adult daughter. Every vehicle the family had ever owned sat rusting in the front yard. Few of them ran.

I first delivered service to Doris when her husband accidentally backed over her with a car. We were all quite astonished to find Doris uninjured after we extricated her from the vehicle's undercarriage. Mr. Doris wasn't so fortunate; after we got Doris out from under the car, dusted her off and determined she was intact, she nailed her husband in the back of the head with one of the many empty bottles littering the front yard.

Doris called us at least once a week, always after midnight. Generally, her daughter greeted us at the front door and offered us tea or malt liquor. We usually found Doris sitting in bed next to her snoring husband. She always complained about "difficulty breathing." She described her symptoms in 30-word sentences while she smoked one of three lit cigarettes. The description concluded with her showing us a water glass filled with phlegm balls. The fact that the family used the hubcaps from their non-functioning vehicles as ashtrays never seemed to convince Doris that smoking might be the cause of her late-night lugies. There were half a dozen of these wheel covers positioned throughout the house. Each held a carton's worth of spent butts, all of them little altars to Doris's nasty sputum and our late-night visits.

Since Doris' cries for help came into our dispatch center as "difficulty breathing," dispatch sent Engine 59, a paramedic engine, as well. The first couple of times, they actually evaluated Doris. The medics reviewed the vitals we had taken, took more of their own, looked into her nasty glass of goo, hooked her up to the heart monitor, and used scientific words to describe her condition. I would say "hack," and they would correct me with "expectorate." Doris loved it up until the time the medics pronounced both her and her heart well enough to swim the English Channel. Then they lectured her about her smoking. The medics would end their happy little visit by telling Doris she didn't need to go to the hospital. We then wished everyone a good night, mounted our beautiful red fire engines and headed back into the random gunfire of the night, anxiously awaiting our next customer-service request.

Not an hour would pass and Doris would call again. Our crew quickly figured out that Doris wouldn't stop calling until she got to show her vile body fluids to "a real doctor like Marcus Welby." Compounding the problem, our alarm room wouldn't stop sending us. We tried to reason with them, to no avail. Their explanation started with liability issues (we could be sued) and ended with our department's policy of sending medical help to people who complain of difficulty breathing (they could be having a heart attack). At one point I was told this policy was set by a medical-control doctor whose brain was so big that his forehead was the size of a 6½-lb. can of hominy. (I was convinced our alarm room got a commission on every call they dispatched.)

We did our little dance with Doris about once a week. There was no sense in having two crews respond, so we would cancel Engine 59, go through our ritual with Doris and wait for the ambulance to arrive. I found ways to occupy my time. Everyone is born with some type of talent, and Doris' daughter had a talent for jigsaw puzzles. It would take our entire crew several months to complete what this savant could do in a week. There were thousands of small, odd-shaped pieces spread out all over the kitchen table, and I took great joy in hiding a single piece from each puzzle. Placing a piece behind a hanging picture or under a table lamp helped me retain a small amount of sanity and kept me out of Doris' bedroom and away from her tumbler of oysters.

☠☠☠☠☠☠☠☠

Doris was just one of dozens of lunatics we saw on a regular basis. This was a constant in our area, and we all knew that. But so were working structure fires, and we hadn't had seen a building burn in two months. Fires were our medicine, and the gods were withholding them from us. Why was it never the other way around?

People have described what we do as selfless and giving. Every time we go and visit Doris, we are those things. We will take hundreds of blood pressures, remain patient during countless encounters with the broken and addicted and perform CPR on the obviously dead in an effort to comfort the family members. We make a huge difference in people's lives, and although it is the most profound thing we do for society, it isn't the driving reason we joined or why we stay. We do it based on the promise that sooner or later we'll be standing in the center of a building that nature is consuming in flame, heat, gas and smoke.

These were the things I hoped for that cold December morning when the alarm room jolted the dorm from dark and quiet to bright and loud. A female voice replaced the tone. She told us a house was on fire just down the road.

Everything seemed to happen very quickly. I looked over at our wide-eyed booter putting his pants on backward. Before I was able to force myself to sit up straight, our engineer was fully dressed and seemed to be halfway through his cigarette. I stepped into my bunkers, pulled them up, laid the suspenders over my shoulders and headed to the rig. The clock on the wall said that it was 0325. We all stumbled out of the hallway that led from the dorm to the apparatus floor, but our engine was not parked in the apparatus bay. We had left it out back after our previous call, a car fire about 45 minutes earlier, to fill up the tank with the "green line" (a technical firefighting term for "garden hose").

As I walked out of the bay to get on the rig, I was assaulted by the smell of the trailer park located behind our firehouse. A month before, one of the single-wide trailers burned when a group of meth-addled bikers tried to cook a large stray dog for Thanksgiving dinner. Since the trailer was so close to the station, the fire was quickly extinguished. The landlord figured the thing could be salvaged, so a couple days later a large dumpster was placed between the burned-out trailer and the vacant trailer next to it. Work didn't progress as planned, and the trailer-park residents started using the dumpster for their own waste. The garbage quickly piled up. As I stepped onto the rig, I could see the still burned-out shell of an aluminum mobile home that had 5–10 feet of trash above its roofline. The mountain of trash was literally pushing the two trailers on either side of the dumpster off their foundations.

The engineer started the rig—a 1976 American La France 1,500 GPM pumper—put it in gear and floored it. This piece of apparatus had a turbocharged V8 diesel engine and a short wheelbase. It performed and handled like a really big go-cart, which is precisely the way our engineer drove it.

It was so cold out that some of the water that had overflowed while filling our tank had turned to ice. This caused the rear wheels to spin when our engineer stomped the gas pedal to the floor. As the rig pushed itself into the apparatus bay, the tires caught and slammed the truck, and all of us, out the front door with a great deal of noise. As the engineer made a sharp left onto the street, we could see the fire's glow a mile or so down the road.

The red sky triggered a release of adrenaline in each of us. The firefighters rode in rear-facing open jump seats. This allowed us to stand up while getting dressed (a practice that is now strictly forbidden). Part of our route to the fire took us across a section of railroad tracks. This portion of the ride required a seatbelt. Both the boot and I were fully turned out and safely buckled into our seats, which were really nothing more than cush-

48

ions bolted to a metal frame. The engineer hit the crest at about 60 mph (he had slowed down for it). I watched 1,000 feet of 3½" hose rise 3 to 4 feet out of and settle back into the rear hosebed. A policeman parked at an adjacent Circle K estimated that our truck was airborne for approximately 75 feet.

Between the first and second touchdowns of our flying circus, I had an overwhelming urge to urinate. I'm sure this was the result of a tight seatbelt and the 6 g's my full bladder had just endured. The engineer later told me the key to this aerial maneuver was not to apply the brakes while in midair. Evidently, upon landing the truck will go into an uncontrollable skid because the brakes are locked. This was very illuminating to me, as it was something I had never thought of before.

Our engineer quickly regained control of our bouncing engine company without losing any speed. The captain of this rolling asylum was pounding on the plastic sliding window that separated the heated front of the cab from the open jump seats where the boot and I sat. I turned my head and saw in the distance the rear of a two-story house that had heavy smoke showing from one of the second-story windows and the gabled ends of the roof. My captain shouted at both of us to get our masks on and pointed straight ahead. I saw Engine 59 screaming down the road from the other direction. About half a mile separated us. I stood up to throw on my SCBA. I looked over to tell the young, nameless booter to follow suit. Evidently, the young man had some modicum of sense left, and it had taken him deep into prayer. I told him to snap out of it and get his bottle on. He complied with my request. We had a three-truck-length lead on the opposing engine company and were preparing to make a right turn onto the side street where our burning house lived. Our engineer swung wide to the left before making his right. This thrust us into Engine 59's trajectory, scaring the opposing engineer (a paramedic, I might add) into slowing down and ending the race. This did two things. First, it ensured we would be first in; second, it caused Engine 59 (the losers this late evening) to lay a supply line for us.

Having survived our response and vanquishing Engine 59, our thoughts quickly turned to the fire. The burning two-story, wood-frame house had been built in the 1940s. Smoke and fire were pushing out of a first-floor window adjacent to the front door. A decent amount of fire was coming from the side of the house just below the roofline. There was also quite a bit of smoke coming from the rear. The young firefighter and I pulled a 1½" attack line as our captain gave an on-scene report and took command. Our engineer was waiting until we got our line stretched to give us water.

Engine 59 did not stage and await orders; they just came in behind us. One of the firefighters was throwing on his bottle while the other one stretched an attack line. Our captain came off the rig screaming at Engine 59's marauding captain. "Get us a supply line! Why didn't you cocksuckers stage?! Blah, blah, blah…" The captain of the other rig responded with some drivel about our engineer playing chicken and endangering the lives of his entire crew. Our engineer flicked his cigarette toward the accusing officer.

In the meantime, the youngster and I had the front door of the house splintered and torn from its hinges and were waiting for water. Heavy black smoke was pushing through the top of the front door as Engine 59's crew was stretching their hoseline toward us. We could see the fire was burning just to the right of the front door, in what appeared to be the kitchen, and was pushing its way toward the front door. Although quite a bit of smoke was venting from the doorway, we didn't feel any heat because most of it was funneling up the stairs to the second floor. The smoke and superheated fire gases rose to the ceiling in the stairway. Every few seconds, these products of combustion would get hot enough to ignite. The flame would move like a slow wave across the ceiling to the second floor landing. If I were Ansel Adams, I would have certainly taken a picture and captured its natural beauty.

As we waited at the door for water, a small fireball shot down the stairs. I had seen fire do some odd things, but I had never seen it run on the ground. Before any of us could figure out what was happening, a burning cat bolted past us and down the street.

I felt the line in my hand twist and heard the rush of air that always precedes the water. When the water hit the end of the hose, it slammed the nozzle forward. We had beat Engine 59 through the front door. The hose was pumped at such high pressure its jacket oozed water. This made advancing the 150 feet of line more difficult, but the high-pressure stream could blast through the crumbling walls and ceilings of the evaporating bungalow. We got through the front door, went to the right and directed our potent stream into the kitchen. The other crew joined us and opened their line also. Within a few seconds, we lost all visibility as the water from our attack lines filled the downstairs with steam. My partner and I went back toward the front door and headed up the stairs while Engine 59 continued to fight the fire on the first floor. As I advanced up the stairs, I started to fall through the burned-out steps (the stairs were above the kitchen, where the fire had been burning for some time). My young partner was behind me, saw what was happening, and blasted me in the ass with his shoulder, which not only kept me from falling through, but also launched me beyond the next few crumbling steps to the second-floor landing. I bounced up, while my counterpart bounced down. The quarter ton of tumbling fire-fighters caused the rickety, burned-out stairs to collapse. The top few steps were all that remained of the stairway. My understudy and I stared at one another, separated by a 6-foot hole of burned-out stairs.

Engine 59's crew came around the corner when they heard all the activity. They knocked down the remainder of the fire in the stairwell and pulled a line for me so I could search the upper floor. The hallway leading to the right was burning, so I went that way. As I moved down the hall it got hotter, and the smoke became thicker. I was soon on my stomach directing the hose stream down the hall. This didn't seem to have any effect on the fire. I couldn't see anything, and the heat was making me very uncomfort-able. I was ready to drag ass when conditions dramatically improved. The fire had burned through the roof, which allowed the heat and smoke to vent. This cleared the hallway and allowed me to stand up and advance my line into a bedroom at the end of the hall. The door was partially open, and the top half of it had been burned off. The entire room was on fire, but you could see everything. The room's contents were framed in darker shades of fire, and everything in the room glowed in the orange consumption.

The bed was the first thing I saw as I stepped into the room. The metal frame radiated a phosphorous white; the mattress springs a brilliant orange. You could see the individual planks that made up the floor beneath the bed. There were only two other distinguishable things in the room. The most notable was a piano against one of the walls. It was entirely intact, but I could see through it. The large instrument's wooden exterior glowed like molten metal. The interior components were outlined in different shades of orange.

The other side of the room contained an easy chair occupied by what I'm sure was a living person a mere 30 minutes earlier. The combustion process had taken hold of him, and most of his skin had been cooked off. In fact, the arms were almost devoid of all tissue. The burning thing no longer looked human—it appeared as if it were made of dissolving clay. The clenched teeth were the only intact remains. The whole scene was quite intriguing. I opened my line on a medium fog stream and briefly directed it toward the burning body. It quickly disappeared in a cloud of steam and smoke. Once the water vapor and smoking suet dissipated, I could see the smoldering human. Both of the arms had been blasted off at the elbows, as if someone had cut a string.

Life is quite odd. Fifteen minutes before, I had been deep in the pleasures of REM sleep, quite comfy under six layers of blankets. Then I was standing in the middle of a burning bedroom, hydraulically disassembling a corpse. I had taken in as much of the moment as I could. Pieces of the walls were starting to follow their burning owner's lead. Good-sized chunks of ceiling rained down, and the roof started to fall apart. The chair the corpse was sitting in disintegrated. I opened the line and started to liberally apply a straight stream of water to the entire room. Steam rose from the floor and walls. The fire only ceased when I applied water. As soon as I moved my stream to a different area, the fire I had just knocked down would reignite. It was like those stupid trick birthday candles you can't blow out. The piano was starting to break apart but wouldn't extinguish. I held the line 6 feet behind the open nozzle and used it like a hand tool to beat the piano to pieces. The stream sent large chunks of the defunct musical instrument flying into the room. I pounded on the piano until it was gone. The room was full of power and a dead guy's intestines cooking on the smoldering floor. Everything was coming apart. Fire and the 200 lbs. of force behind the water I controlled were fucking shit up. I was transformed. A guy can get lost in himself during these types of life experiences. Doris no longer existed. This little psycho waltz lasted no more than 4 or 5 minutes. I had reduced the fire from a deadly menace to an occasional flicker. I looked up as I dropped the line. Most of the ceiling and roof had burned away. I could see the moon.

Chapter 5
The Shifts

Most organizations with around-the-clock staffing have three eight-hour shifts. Fire departments don't. Our ancients chose to work 24-hour shifts. Instead of having morning, swing and graveyard shifts, we have A, B and C shifts. If you were to visit a fire station every day for a week, you might not notice much difference between the shifts assigned to that station. To the casual observer, firefighters pretty much look and act the same. This changes dramatically once you join the organization and get categorized as an A-, B- or C-shifter. The shifts' personalities are at the core of all of our stereotypes and are also the stuff of legend.

The first day of the year begins on A Shift. A Shift is the first thing God created. A Shift is No. 1, and its members display typical first-child syndrome. They are the bright, bold and beautiful backbone of our organization. A-shifters are the go-getters—upwardly mobile overachievers who love to blow their own horns. They are adept at taking credit and shifting blame. They are trendsetters with impeccable hygiene and expensive haircuts. An A-shifter invented dental floss. They are so sparkly your retinas would explode if you looked directly at them.

A-shifters are a tight bunch. When A-shifters flock, they like to use words like "stud," "athlete" and "Viking warrior" to describe one another. When they really admire a fellow A-shifter, they'll say things such as, "I'd storm the gates of hell with him. He's an outstanding fireman." If you bounce this needle of male admiration just a click or two south, it registers as fellatio (a Latin word meaning "to suck the cock" for all you fellow B-shifters reading along).

A-shifters tend to gravitate toward jobs at the training academy when they promote to captain. Most A Shift training captains will tell you the reason they love to train recruit firefighters is it allows them to give back to the department. Don't believe that bullshit for a minute. A-shifters love training new firefighters because it provides them with the ultimate platform to show off and lord over the department's newest members. It's the same personality defect that causes the high-school track coach to fuck all the pretty junior girls.

Next comes B Shift. The rest of the organization regards us as a bunch of fugitives. An A Shift chief once said he would rather have a sister in a whorehouse than a brother on B Shift. (Several years later, he was transferred to B Shift. On his first day, he received dozens of phone calls from loving B-shifters welcoming him aboard and asking about his sister's welfare, whereabouts and favorite position.) B-shifters are fond of eating with their hands and breaking things. A few years ago, someone in our department did a study regarding our discipline process. The study indicated that 70 percent of all personnel problems occurred on B Shift. This is a pretty accurate testament to B Shift's philosophy regarding rules.

C-shifters strive to be offended. They have the innate ability to stare into pure joy and find the shadow of dark despair. Nothing is ever clean enough or orderly enough in the flawed world that surrounds them. C-shifters invented both the form and the sponge. They are natural accountants, supply clerks and janitors. From across the room, they can tell you when the stove was last cleaned. C-shifters feel more at ease when they herd together in large groups. This led a prehistoric tribe of C-shifters to invent the meeting.

A-shifters consider their bodies a temple, B-shifters consider theirs a tool and C-shifters are afraid of theirs. A Shift likes to make the rules and study their profiles and chiseled abs in the mirror. C Shift loves to follow rules and whine. B Shift loves to self-actualize and blow stuff up.

In truth, the differences among the three shifts are minuscule, but like all other stereotypes they provide our inefficient human brains little handles it can hold onto. I have worked with C-shifters who were not only excellent firefighters but also practiced hedonism at a level that would kill a B Shift truck company. Some of the toughest and most humble firefighters I've had the pleasure to work with were A-shifters. I have also known B-shifters who were clean freaks and recited poetry. Stereotyping individual firefighters by shift creates sub-cults within our already secretive fire-department cult. This is okay when it allows us to tease one another. On the other hand, it's something the elders of the department must watch and manage to ensure it doesn't get out of hand and turn evil. After all, it was a group of sub-cultists who created what would become the Nazi Party (probably a bunch of A-shifters).

☠☠☠☠☠☠☠☠

The fire station kitty is one of the things that magnifies the differences among the shifts. Running a station kitty is a thankless job. Stocking a fire station with peanut butter, coffee, spices and other sundry items along with paying the monthly cable and newspaper bill is time-consuming drudgery.

The kitty man rarely receives so much as a thank you. Most kitty-related comments are negative and take the form of sniveling and whining about what the kitty does or doesn't have. "The kitty at Station 6 buys hair conditioner and blowdryers; we just have shampoo. Our kitty sucks."

Another kitty-related pain in the ass is collecting the money. I have seen firefighters spend hundreds of dollars in a saloon the night before work only to hear them bitch the next morning about having to drop their bimonthly $4 for kitty. "I don't drink coffee or read the paper so I should only have to drop $2." Any kitty man worth his salt will quickly humiliate these cheapskates and shame them into paying their full kitty fee.

I began my career as a firefighter assigned to a south-side firehouse. When I started at this station, a kitty war was playing out between A Shift and C Shift. The kitty hostilities began when Marcus, an A-shifter and the station's kitty man, started buying bottled water for the station. It had something to do with our chunky tap water and its paint-thinner smell. The water in Phoenix has so many minerals in it that people have chipped teeth drinking it. C Shift, a bunch of big, tough cowboy types, thought bottled water was only for office workers and sissies. But what really pissed off C Shift was the absence of a vote. Before the kitty makes any large or unusual purchases, there's supposed to be a vote among all three shifts. C Shift maintained that these were the ancient customs. A Shift told them to kiss their ass and continued to buy the things they wanted. C Shift got revenge by secretly dumping the water out of the 5-gallon jugs and refilling them with the garden hose. A Shift caught on two months later when they noticed grass floating on top of the water.

My own trouble with the kitty Nazis started one day when I came to work and noticed three sticks planted in the front yard. They looked like tree trimmings that had fallen off a truck heading to the dump just south of our station. I asked Marcus what they were. He proudly told me they were fruit trees purchased with kitty funds. I asked him if C Shift got a chance to vote on them. He told me to kiss his ass and left. Feeling good that I was able to upset one of the A-shifters before he went home, I went into the kitchen to scrounge for some of their leftovers. The refrigerator was as bare as Old Mother Hubbard's cupboard (A Shift didn't believe in excess). I made my way to the freezer and bingo, homemade ice cream. A Shift made ice cream almost every shift. Ours was the only kitty in town stocked with 50-lb. sacks of sugar and cases of evaporated milk. I pulled the ice cream out of the freezer and removed the masking-tape barrier that read "Save— A Shift" and opened the lid to find more than a quart of delicious strawberry ice cream.

Saving leftovers is one of our ancient customs. Sometimes there will be enough dinner leftover to serve for lunch the next shift. The crew wraps the leftovers and marks them with the word "Save." We are pretty good at not violating the sanctity of saved food items, but A Shift had no business saving ice cream I helped finance through my kitty fees. I sat down to enjoy my cold and creamy breakfast and made a little promise to myself that I would make a practice of this indulgence. For a couple weeks, I gleefully consumed A Shift's leftover dessert treats. One morning, Jeff, the A Shift captain, mentioned this to my captain. "Larry, tell your animals to stay out of our leftover ice cream." Chuck, our B Shift engineer, piped in. "Jeff, kiss my ass." B Shift frequently communicated with the other two shifts in this fashion.

After Jeff left, Chuck and I shared the last of A Shift's peach ice cream. Larry tried to stop us. Chuck looked at Captain Larry and said, "Kiss my ass, you stupid Pollack." (Chuck and Larry had worked together a long time. Their relationship was more like an old married couple's than the traditional boss-worker relationship.) I informed Captain Larry this was actually kitty ice cream, and, that being the case, it was really all of our ice cream. After Chuck and I finished our morning snack, we went out front to gaze upon the blight that was our first-due area. I noticed one of the sticks A Shift purchased had sprouted some leaves. Chuck and I quickly decided there was one tree for each shift, and we were going to decorate ours.

About a month previous to our tree-trimming extravaganza, we had flunked our monthly station inspection. Our chief, who was a real stickler for cleanliness, did not like the condition of the air vents and the fuel pump. He ordered us to clean them up and paint them. We had several cases of spray paint. The problem was we only had two colors—red and silver. I called our Building and Grounds Department to inquire about getting a color that matched the station. They informed me I would have to fill out form 97-61 and forward it to my battalion office, which would then route it through the proper channels, and I could expect my paint in anywhere from 6 months to 87 years. I resolved to go with the red and silver. Red was a natural color for the fuel pump. The silver barber's stripe I used to accent it was not. I went crazy. I painted the vents red and silver. Next were all the telephones. They stopped me when I started to do the kitchen cabinets. Chief White Gloves wasn't too happy, but at least it was fresh paint.

Chuck and I went to work on the tree. We painted the top half silver and the trunk red. It really livened the place up. A Shift was not pleased with our decorating endeavor. I tried to explain I had painted B Shift's tree. They told me those were the kitty's trees, not just B Shift's. Chuck interjected that if that was the case, they were all of our trees, and we would paint the other two. Jeff said not to paint anything else. He added he had taken actions to ensure we would stop eating their leftover ice cream. I did not like the way he smiled when he left the station.

All the silver leaves fell off our tree, and Kitty Man Marcus came out front to yell at us for killing it. He had a lazy eye, so it was hard to tell if he was directing his comments to Chuck or me. I told Marcus the tree was just in a minor form of shock brought on by being so beautiful. Marcus stormed off again. After he left, Chuck and I spruced up our tree by taping palm fronds and weeds to its bare silver branches.

About an hour later, Chief White Gloves showed up. I thought he was there to drop off the mail or to inspect something. The only talent I ever saw this man exhibit was the ability to judge the cleanliness of different objects, and he really did have a pair of white gloves. More complex issues, like burning buildings and mass-casualty incidents, seemed to confuse him. He sat the entire crew down to talk about some pressing matter that had him sweating. He told us A Shift had filed a police report regarding a stolen item. The reason for his visit was to investigate what might have happened to the missing "thing." Larry asked him what was missing. White Gloves, who looked like a badger was trying to eat its way out of his chest cavity, said "ice cream." The man was clearly in over his head. I tried to tell him that it was kitty ice cream and no one could claim it as their own. He cut me off as he lunged from his chair and started to rant about the seriousness of these allegations. Before I could reply, Chuck shouted, "That's it!" He pointed at me and said, "Don't say another word. Chief, if this is the case, we refuse to speak until our attorney is present. If you don't have any other business I ask that this meeting be adjourned." I seconded the motion. Chief White Gloves collapsed back into his chair. As he sat there mumbling, I asked if he'd been out front and seen our pretty little tree.

☠☠☠☠☠☠☠☠☠

Our department has a lot of rovers. When a member begins their career or is freshly promoted, they typically will not have a permanent station. They start their shift by calling in for that day's assignment. They repeat this routine until they accumulate enough seniority to get a spot. Rovers work all over the city and get a diverse look at the mass of humanity that makes up the department—including the wide variety of personalities on all three shifts. When I was a newly promoted captain, I roved into Ladder 81. Station 81 serviced an active part of Phoenix (the best late-night marksmen in the city resided in Station 81's first-due area). The Ladder 81's B Shift crew was very competent. They provided excellent customer service on calls, kept the station and apparatus squeaky clean and enjoyed having fun while they did their jobs. The A- and C-shifters were the same. You would think this would foster harmony among the three shifts. Think again.

As a rule, C-shifters come to work looking for things that are out of place or not done according to their standards. One morning at shift change, I was talking to Paul, the C Shift ladder captain. Hector, one of the C Shift engineers, interrupted our conversation with a piece of urgent news. He was holding a lug nut off one of the rear wheels of the ladder truck. "Look at this, Skip. I found this loose lug on the right rear dual of the rig. B Shift never checks anything. This should have been fixed yesterday."

Paul gave me a disappointed look and said, "Nick, while you're here I would appreciate it if you made sure your guys did their jobs and took care of these things."

I started to respond when Bob, one of the B Shift engineers, spoke up. "I took the rig in yesterday to get one of the rear tires changed; it looked like all of the lugs were on tight when the mechanic finished. I don't know what your routine is, but ours doesn't include checking the lugs every hour."

Hector was not going to let this go. His sense of order had been violated by the wayward lug nut. "This is bullshit. We are always finding stuff like this. Last shift the rig only had half a tank of fuel when we came on. The shift before that, the siren didn't work right. I could go on but it doesn't seem to do any good." I was in the middle of telling Hector not to worry about it, there were another 27 lug nuts holding the wheel in place, when Wayne, the other B Shift ladder engineer, made his presence known. He had been sitting quietly at the end of the table eating a bowl of Cheerios and reading the newspaper. He sprayed milk out of his mouth when he shouted, "Do you guys know why we never do anything?"

Everyone remained silent, waiting for him to answer his own question. He looked around the room at each one of us and said, "It's because we're relieved every morning by 10 angry cleaning ladies."

☠☠☠☠☠☠☠☠

The different shifts' personalities and relationships with one another have existed long before I got here and will continue long after I'm gone. When I was just starting my career, an older member explained the differences between the shifts with an event that happened in the 1950s. The chief had just awarded a safe-driving watch to Engineer J.W. Robinson for five years of accident-free driving. J.W. was the first black man hired by the Phoenix Fire Department. He was very hard working, tough, smart and proud. He was a typical 1950s era A-shifter. The chief assembled all the men of the multi-company station and presented J.W. his watch. In typical A Shift fashion, the men made a big deal of J.W.'s driving prowess as they told stories and stretched their scrotums.

J.W. drove the hose truck. Dago Don, his counterpart, drove the engine out of the same station. Don never got a safe-driving award. Don may have been assigned to A Shift, but he was a B-shifter through and through. Don's hobby was racing sprint cars at the local racetrack. Many of his races ended in an accident. This happened often enough that Don's mechanic (who was also a firefighter) told him, "Son, if you don't quit using your head as a roll bar, the city's going to make you become a cop." Most of Don's races ended with an ambulance ride to the hospital. This happened with such regularity that Don's mother became very concerned about her son's life expectancy. Her house was located between the track and the hospital, so every time the ambulance hauled Don to the hospital, they first stopped at his mom's house. They would blow their siren in the driveway and mom would come out. Don would sit up and wave at her through the back window of the ambo before embarking to the hospital for his weekly neurological exam.

A few hours after J.W. received his safe-driving watch, the entire station was dispatched to a structure fire. For some reason, the hose truck J.W. was driving spun out of control and rolled over. Don was following behind J.W. when the accident occurred. Don's captain and crew were very worried that J.W. had been seriously injured. Don pulled over and sprinted to J.W. and the upside-down hose wagon. The rest of the crew followed close behind him. Don got on his hands and knees, reached in through the rig's window, grabbed J.W. by the wrist, and started pulling. Don's captain saw the accident had knocked J.W. senseless and told Don to calm down, he didn't want Don to aggravate any injuries J.W. may have suffered. Don kept grabbing and tugging anyway. J.W. was starting to come around from all of the shaking. When he was finally clear-headed enough to talk, he asked Don, "What the hell are you doing?" Don replied, "Son, I'm getting the watch back for the chief."

☠☠☠☠☠☠☠☠

I was assigned to Station 66, where I had the pleasure of managing a long line of probationary firefighters. When firefighters graduated from the training academy, they rotated among three different training stations during their first year. Most of this nameless organizational youth came and went, but occasionally we would get one we didn't want to give up. My favorite was a young lad named Dennis. We were Dennis' first station out of the academy. The morning of his first shift, he introduced himself to all of us. Being one of his captains, I sat down and told him the station routine. He was very quiet and polite. He told me he was looking forward to the next three months he would be spending with us. I soon discovered I was very wrong to categorize Dennis as just another probationary firefighter. I would find out over the next three months how special our little B-shifter was.

After finishing the day's lunch, we were cleaning the kitchen and learned we were out of dishwasher detergent. Jake, one of the engineers, decided Spic and Span looked a lot like dishwasher soap. He filled up our low-bid dishwashing machine and turned it on. I retired to my office to catch up on paperwork and other silly organizational chores. The TV was on. A movie with Robert Blake was playing. It garnered my attention when the former Little Rascal started having sex with a young retarded boy. The two actors were cuddling after they consummated their newfound relationship. Robert Blake was stroking the dim wit's hair and kept referring to him as his "little human." Definitely a couple of C-shifters. I was pulled away from this love story by a knock at my door. It was young Dennis. He came to tell me the dishwasher had pumped a 4-foot high pile of suds into the kitchen. I went out to the kitchen to find a river of bubbles running the full length of the kitchen and heading down the hallway toward the dorm. I told young Dennis not to worry about it, it would evaporate before dinner, but to call the red shirt (a fire department delivery boy) and have him bring us some real dishwasher detergent. Dennis said he'd he take care of it. I went back to the cozy confines of my office and was followed by our newest member. Dennis asked if he could sit down and talk to me about something.

"How long have you been a captain, sir?"
"About six years."
"Do you like it?"
"Yes."

60

"Do all captains get their own office?"

"Yes."

"I think I would like if it I was a captain and got my own office, sir."

"Well Dennis, you'll be eligible to take the test in seven years."

"Sir, do you ever spank your monkey in this office?"

"Excuse me?"

"You know sir, do you ever masturbate in your office? I think that would be the best part of being a captain and getting your own office. I have to sleep in the dorm with eight other guys, and I think if I started to punch my clown in there some of them might hurt me. In fact, there are some guys working here who look like they're capable of raping me. I'm a little scared, sir."

"Don't be scared. I'll take care of you, my little human."

Our next shift began with a kitchen full of dirty dishes. It looked as if A Shift had catered an Italian wedding the night before. The group of us was cleaning the kitchen when we discovered we were still out of dishwasher detergent. I asked Dennis if he had any luck getting some delivered last shift. He told me the red shirt hadn't been able to drop any off. The last off-going A-shifter wandered into the kitchen and informed us the red shirt delivered the stuff on A Shift morning, but C Shift had sent it back. "C Shift was pissed off you guys ordered supplies because that's their job, so they ordered the red shirt to take it back to the warehouse."

"Well, that's swell. How are we supposed to do dishes?"

Our engineer Jake put the newspaper down long enough to say, "Spic and Span."

C Shift is its own worst enemy. They need to live in an orderly universe. They order the supplies. Forget that we're out of something and need it. The rules clearly state C Shift is responsible for ordering commodities. They brought the following situation on themselves.

The off-going shift will generally clean the kitchen before they leave. This is only fair. We eat the evening's leftovers for breakfast, have coffee and make a mess of things. That morning, the last B-shifter to leave filled the dishwasher, loaded it with extra foamy Spic and Span, shut the door, turned it on and left. About 20 minutes later, the kitchen was full of tiny bubbles, and C Shift was fit to be tied.

The next shift change, one of the C Shift captains approached me about the dastardly deed we had committed. I told him we were sorry but it was out of habit, not spite. The captain replied, "C Shift orders supplies. If you guys need something let us know, and we will order it. We don't like you ordering things. In fact we don't like putting up with a lot of the shit you guys pull."

"The three shifts don't do things that differently from one another," I replied. "I've never noticed the world falling apart when B Shift is on duty."

"Yeah, if that's the case, how do you explain the fact that last month you assholes gave away all of our station dishes. We had to borrow plates and silverware from Station 88. The next shift we came in only to find you guys gave all those away, too. We called the guys at 88 to ask for another loan, but you morons had already broken into their station and stole all their remaining dishes. If you idiots don't mind eating right off the table that's your business, but I'm sick of having to live with the consequences of B Shift's juvenile antics."

"Don't confuse yourself," I said. "We go the extra mile on the B Shift. B Shift has taken it upon ourselves to expand our service delivery. The property owners in this area can't afford nice things. If we go on a call and find the customer has substandard dishes, we help them out. Don't you remember when you were a kid and your parents received a free place setting with every tank of gas? Didn't that make your mommy happy? We're giving back to the community. We provide top-flight EMS and complimentary place settings. Our customers love it. Besides, the guys at 88 are a bunch of prima-donnas and douche bags." His eyes seemed to tear up before he slammed the door on his way out of my office.

After Captain Grumpy Britches stormed off, I changed into my civilian clothes and was in the process of leaving when a minor argument broke out in the kitchen. One of the C-shifters was yelling at one of the B-shifters because the dishwasher was generating a large amount of suds. The arguing duo was standing in the middle of 4 feet of white, soapy bubbles that filled the narrow width of the kitchen. I screamed at the B-shifter to shut up and go home. I told the whining C-shifter, "Shut up and clean up all these goddamn bubbles you assholes made us make. If you don't quit complaining, I'll start giving away the pots and pans on calls."

The next shift change between B and C Shift played out the same way. We can all be so petty. B-shifters were now dissolving Spic and Span in the bottoms of coffee mugs. Unsuspecting C-shifters would load the mugs into the dishwasher without dumping their contents into the sink. The final straw was when an anonymous B-shifter (probably Dennis) dumped the dishwasher detergent out of the 10 boxes C Shift had finally procured and filled them with Spic and Span. This happened on Halloween. That morning, C Shift had purchased a metric ton of candy to pass out to the trick-or-treaters and local junkies. Pookie, one of the C Shift engineers on the ladder, was in charge of passing out the silo of goodies. Pookie ate three pieces for every one he handed out. One of the B-shifters was working an extra shift with his C Shift brothers and sisters. At about 10 that night, Pookie started to complain of chest pain. He was in his 50s, so the C-shifters figured he was having the big one. The Engine 66 medics started an IV and put the heart monitor on him. His heart rhythm indicated his ticker was fine, but they took him to the hospital just in case.

The next shift we were attacked by C Shift. They accused us of poisoning Pookie with whatever foaming solution we were putting into the dishwasher. Mike, the B-shifter who watched Pookie eat his weight in taffy, spoke up.

"Bullshit. We didn't poison anyone. Pookie ate enough candy to kill a diabetic giraffe. He didn't have a heart attack, and he wasn't poisoned. He ate his weight in candy and got an upset tum-tum."

Pookie sat there with big crybaby eyes while a female C-shifter rubbed his shoulders and consoled him like a child. One of the C Shift captains was not buying it. We had disrupted his order.

"We've had it with your shit. You guys don't do anything around here but figure out ways to fuck with us. It's going to stop."

I attempted to make peace with the C-shifters. I explained the dish-washer fiasco began as an innocent mistake that snowballed out of control. Both shifts were equally responsible. They were having none of it. One of the more vocal C Shift firefighters felt we had corrupted young Dennis into a B Shift joker. "That kid was borderline when he showed up here. I wonder what rocket scientist at the training academy thought it was a good idea to pair him with you assholes. The most endearing thing he's done for his reputation was super gluing his pubic hair to the urinal. You've turned him into a first class B-shifter."

The C-shifters were right. Dennis was a dyed-in-the-wool B-shifter, our poster child if you will. We loved him.

The hostilities pretty much calmed down for the rest of Dennis' tour with us. We were all very sad when young Dennis had to leave. His last shift with us started out at the training academy, where we attended a class on infectious-disease control. Our EMS Division passed out HEPA face masks—the latest and greatest in hepatitis protection. These masks were made of some type of thick purple germ-fighting material. They looked more like dog muzzles than a piece of high-tech EMS gear.

We all received our special masks and were sent back into the city to fight viruses and the like. When we got back to the station, we fixed lunch. As we sat down to eat, we noticed Dennis wasn't at the table. One of us used the intercom to tell him lunch was being served. Dennis emerged naked, wearing only his HEPA mask like a fig leaf. He jumped up on the table and did a little kootchie-koo dance. We all felt safe knowing we couldn't catch hepatitis from Dennis' thingy.

C Shift classified Dennis as a demented little B-shifter—a bad example for future generations—and celebrated his departure. They would not have used a word like "hero" to describe him. It wouldn't be the first time they were wrong. About five years after my little human left us, our fire chief would hang our department's Medal of Valor around his neck.

One day Dennis was driving with a friend of his who was trying to get hired by our fun-filled organization. Dennis saw a column of smoke a few blocks away and figured it would do his young associate some good to watch us operate up close and personal. He and his young charge drove toward the smoke and made it to the general vicinity of the fire.

From his vantage point, it appeared something behind a large Wal-Mart shopping center was on fire. He drove behind the store, expecting to find an industrial-size dumpster burning near the rear of the building. When Dennis got to the rear of the structure, he saw the fire was in a three-story apartment complex directly behind the shopping center. Dennis got out of his truck and jumped the 6-foot block wall that separated the store's parking lot from the apartments. He determined the source of the smoke was a third-floor apartment. Dennis was on the backside of the units and could see smoke rolling out of the balcony. All of the neighbors were starting to congregate. The fire department wasn't on the scene yet, so Dennis instructed one of the bystanders to call 911. Flame started rolling out of the arcadia door, and a large man wearing a robe appeared on the balcony. Dennis yelled at the trapped tenant to get down. The fire had the man very confused and scared. It took several minutes of screaming for the man to retreat under the smoke and heat. You never know how people are going to react when faced with serious peril. Some have to be told the obvious repeatedly, and they still don't understand. Other people will respond to their own crisis like it's an everyday event.

Dennis ran around front to see if he could make entry into the fire apartment through the front door. He climbed the stairs to the third floor but encountered too much heat to get inside. He made his way back to the rear of the apartment. The large robed man was now on his hands and knees at the balcony railing, trying to avoid the scorching heat and flames. Dennis yelled for the man to climb over his railing and lower himself to the second-floor balcony. The poor bastard couldn't move. He was paralyzed with fear.

Dennis' father once told me the high school principal would occasionally call him to report Dennis had climbed a tree and was refusing to come down. Dennis Sr. told the principal not to worry about it; Dennis didn't like the dark, and he would most certainly come down before nightfall. I suppose the main reason we go to school is to learn. At the core of all this learning, we can discover how we can best serve society. Dennis spent a good amount of his educational time learning how to climb. The C-shifter in him might have felt this was a waste of valuable school time. The B-shifter voice that spoke to Dennis told him climbing was a noble and worthy activity to pursue. The robed man, who had eaten too many Twinkies and who was about to be burned alive, owes his life to Dennis' misbegotten youth.

Dennis quickly scaled up to the second-floor balcony. He stood on the railing of the second floor and came face to face with the robed man. The fat man's eyes were bloodshot and watering, snot was running out of his nose, and he was panting like a dying animal. Flame was now blowtorching across the patio ceiling. Heavy, pressurized black smoke was pushing out of the melted arcadia door. Dennis looked down to the second-floor balcony to see another Good Samaritan had joined him. Dennis turned his attention back to the robed man. He told him he had to climb over his railing to escape the fire; it was his only avenue of escape. The robed man's eyes were glossed over and empty. Dennis grabbed him by the ear and shouted, "You're going to die if you stay here; climb over the railing, and I'll help you down." The robed man responded with animal noises. Dennis looked back down to the second-floor balcony and saw a heavy nylon cargo strap. He ordered the kid who joined him on the second-floor balcony to give it to him. Dennis wrapped the strap around the railing of the third-floor balcony and threw the other end to the kid on the second floor. He grabbed the robed man by the ear again and told him to get up and put his leg over the railing. Dennis was now standing on the outside of the third-floor balcony with fire burning less than 2 feet above his head. He grabbed the robed man and forced him to get up and put his leg over the railing. Dennis held onto the railing with his left hand and grabbed the robed man with his right arm. He lowered him down to the kid on the second floor. The robed man weighed 100 lbs. more than Dennis.

This entire event took place in less than 5 minutes. Engine 15 had arrived to the scene and began knocking down the fire. Dennis and the kid walked the robed man to the engine company where Dennis gave the robed man a breathing treatment for his smoke inhalation. Dennis was getting ready to take off when one of the firefighters noticed him. He asked what he was doing. The kid who helped save the robed man told the firefighter the events that preceded their arrival. The A Shift firefighter smiled at

Dennis and said, "I can't believe it Dennis, you've gone from zero to hero. You keep this shit up and you'll end up on A Shift."

Chapter 6
My Master & His Pet Sloth

There are many good human traits. Being smart, pretty, clean and multi-lingual are admirable qualities, but if you want to survive for the long haul, you had better be resilient. No one taught me this lesson better than Captain Maynard.

I was a four-year firefighter working at a single engine company in a south side station. This was home to Engine 55, a basic life support (BLS) engine company. We did basic first aid—no IV or drug therapy, no heart monitor and no trips to the hospital to deliver patients. I worked with a captain who liked to run in 10K races, an engineer who had a tendency to run into other vehicles, and a never-ending string of probationary fire-fighters who were assigned to us in three-month increments. The four of us protected the part of Phoenix where one would insert the nozzle if they were giving the city an enema. The glamorous businesses in our district included pallet manufacturers, mulch operations, the county jail, a sewage treatment facility, the town dump, animal-rendering plants and the dog pound. At night, the animal-control people burned the pelts off the animals that had been euthanized during the day. After dark, our area was lit with the burning gas of industry and death.

Our station was built in the 1950s and was identical to a dozen other stations built between World War II and the Vietnam War. It was made of red brick and had a flagpole in the front yard. Inside, the station had a small kitchen and dining room that spilled into the day room. The back of the station included a captain's office and the dorm—a large bedroom that held four beds and a pool table. A bathroom and locker room separated the sleeping quarters from the kitchen and dayroom. If the station didn't have a large two-door apparatus bay attached to its north end, it would have looked like any house on Beaver Cleaver's block.

There were a large number of medical calls in our area, and the powers that be decided Engine 55 needed to have magical advanced life support (ALS) capabilities. ALS meant paramedics, the gifted Children of God who performed miracles. The rank of paramedic was an assignment. Firefighters, engineers and captains were all eligible to take the medic's test. Once accepted into the paramedic program, you went through six months of intense medical training. When you graduated, you received a sizable pay raise, got to have sex with fabulous nurses and could use really big words. You got to wear a special silk-screened medical patch on the right arm of your T-shirt. Many of the medics thought this patch was actually the thumbprint of Jesus Christ himself.

The mystical conversion from BLS to ALS meant two of us were going to have to hit the road. None of us was happy about it, but who were we to stand in the way of progress? After the grand poobahs figured out the complex seniority rule, it was determined the captain and engineer positions would be converted to ALS. This meant I got to stay and face an uncertain future. Rumors were circulating that an older captain who just completed paramedic training was coming to Engine 55, of all places. I didn't want to believe it. It couldn't be true. What would possess a man who was on the verge of retirement to 1) become a paramedic and 2) come to Station 55 and work with a degenerate like me? The fact that he was known for being a world-renowned prick didn't help matters.

The fateful day finally arrived. I got to work and put my turnouts on the rig. I was checking out the EMS gear when I first saw him. He was walking through the open bay door. The sun was pouring in behind him, so I had to squint. I stood to greet him, but I still couldn't make him out because of the glare. I didn't see his features until he was almost right in front of me. The first thing I noticed were the 40 or so small scabs on his face. It looked like someone had taken a knife and made a bunch of little lacerations all over his head, face and neck. We shook hands and he introduced himself as Captain Maynard. "Nice to meet you," I responded. "I'm Firefighter Nick. What happened to your face?" He glared at me and said, "Not that it's any of your business, but I had a bunch of skin cancers burned off." We had gotten off on the wrong foot. The situation was only made worse by our engineer's replacement, a paramedic-engineer who never seemed to like his nickname, "The Three-Legged Sloth." He was a long time buddy of Captain Maynard, much like the Wicked Witch and the Colonel of the Flying Monkey Squadron. My resiliency was about to be tested.

After we checked out the rig and cleaned the station, Captain Maynard sat us down to explain how things were going to be. This was a short conversation that basically centered on the concept that he would be making most of the decisions. He also didn't like some of the pictures hanging in my locker. A few months earlier a group of the local school kids were touring our station and my locker had been inadvertently left open. One of the children pointed at a picture and said, "Look everyone, it's mommy." The problem was that mommy had a mouthful of a large naked man wearing a cowboy hat. Somehow this tale got back to Captain Maynard, and he was not going to stand for this on his watch.

I was used to the way our old crew operated. We showed up for work and everyone did their job. There was not a whole lot of order-giving by our previous captain. Captain Maynard did not subscribe to this theory. He enjoyed giving orders. There was never any doubt that he was the man with a plan. He also had a very dim view of our fire department's senior management. This created a small problem because my father was the fire chief. I'm sure Captain Maynard tempered some of his comments, but it was clear he did not approve of the good relationship that existed between the chief's office and the union. Nor did he like the laid-back management style the chief encouraged. Management's position was that we were going to be nice to the customers, deliver the best service we could, not kill or injure the work force, and have fun while doing our jobs. The heretics. Both Captain Maynard and the Sloth felt this was all too huggy-feely, and that was no way to run an important paramilitary organization like a fire department. Forced marches with lots of saluting were more their style.

One Sunday morning, I came to work in need of some rest. Ancient customs dictated that Sunday naps were appropriate. I came in, put my gear on the rig, helped clean the station, and then I made my rack. I set the a/c to 60 and went to sleep. I was just entering the slobbering phase when someone shook me awake. It was Captain Maynard. He wore an ear-to-ear grin across his thin, Grinch-like lips. He told me to get on the truck. We were going to spend the rest of the day twisting hydrants. He chuckled as he walked out of the dorm. I did as he commanded and spent the next five hours in 110-degree heat. He and the Sloth sat in the cab of the rig while the booter and I went from one plug to the next. They both wore their stupid "we're superior" grins. I vowed to do all I could to make their lives miserable.

After that evening's dinner, we had a call to the county jail. We arrived to the scene and were led through the endless maze of locking doors. The final door dumped us into a large dining room. At the far wall of the room stood a 3-foot-high block wall topped with large sheets of bulletproof glass that reached to the ceiling. The prison cells were on the other side of the glass. As soon as we walked into the room, a bunch of female inmates ran to the glass wall and started hooting and making lewd comments. Captain Maynard and the Sloth looked away indignantly. The booter stared, wide-eyed. I smiled and waved to the girls. They reciprocated by taking off their shirts and pressing their naked breasts against the glass. I could hear the guards laughing as I sprinted to the window. Several of the girls started to play with one another in a very serious way. The girls were going all out and putting on quite a show as I stood inches away, separated by a mere sheet of glass. As I stood enthralled by the magnificent sights, Captain Maynard and the Sloth grabbed me and tried to pull me away. I was too strong for them. I asked one of the guards to let me in. He told me show time was over. Several of the guards made access to my special princesses and escorted them back to their cells, ending our little love fiesta. The old guard smiled at me and said, "Son, they would have killed you." I just smiled at purple-faced Captain Maynard.

The guards led us to the men's side of the jail, which was laid out exactly like the women's side. We went through the multiple set of locking doors and proceeded to the infirmary where one of the prisoners was complaining of chest pain. A group of male prisoners was playing cards around a table. I was saying a quiet little prayer that there weren't 25 guys shoving their peckers against the glass when somebody screamed my name. I looked up and saw a high-school buddy clad in a bright orange prison jumper. I smiled and waved to him. I had to laugh when the Sloth said, "You've been on the wrong side of the glass since we've been here."

The chest pain turned out to be indigestion. When we got back to the station, Captain Maynard took me into his office to scold me. He was quite surprised when I agreed with him and promised I would never look at lesbian sex acts again. I told him the episode had left me feeling dirty. I added that I needed a change and was considering taking the paramedic test. He thought that was a good idea. He droned on for the next 30 minutes about his recent paramedic-training experience. He got to vent about how screwed up the entire paramedic-training process was and the ways he could fix it. As I sat there listening intently to his babbling drivel, I quietly daydreamed of throwing him into a pit full of flesh-eating lesbians. That concluded our meeting.

☠☠☠☠☠☠☠☠

Captain Maynard was big on uniforms. He missed the old fire department get up. In the late 1970s, we transitioned from wearing bow ties and button-up dress shirts with patches on the sleeves to T-shirts. We were the first department in the country to make this change. The biggest problem with our old uniform was that it made us look like cops. When we started delivering EMS to the community, some guys started to wear an EMS pack on their belts. From a distance or at night, it could be mistaken for a gun. The community (especially the one I worked in) has a different reaction when the police show up. They will sometimes shoot at the law, whereas they are usually happy to see us. I liked our uniform even more when I found out Captain Maynard didn't care for it.

I spent my two days off shopping for the perfect pair of sunglasses. They had very large, square black lenses set in a tortoiseshell frame adorned with a dazzling array of rhinestones. The wide metallic sides made a dramatic arc before slipping over my ears. They looked like something Barbara Streisand's mother would have worn. I could hardly wait to get to work.

Despite Captain Maynard's misgivings, I wore my sunglasses all the time. They pissed him off and made me feel special, fueling my resilience. But the Sloth was really starting to get on my nerves. He had been bullying the booter, just because he could. Booters have less than a year on the department and are on probation, which really is the organization's way of saying "we can fire you without cause." They had no civil service review. Some of our more dildonic members thought that gave them an open ticket to fuck with the boot. The Sloth was in the middle of one of his booter-directed tirades about some petty issue. After the Sloth finished, I told the boot not to pay any attention to him, in fact he should take it as a sideways compliment. Both the boot and the Sloth gave me a bewildered stare. I explained to the boot that the Sloth was gay and that this was his way of showing affection.

One Easter night we had a late call for an ill man. We arrived to the scene of a long row of dilapidated shacks. They fronted a very busy road that carried a lot of tractor-trailer traffic. The sign out front advertised these units as "cottages." These were in fact love nests gay men could rent by the hour. They piped gay porno into all the rooms. The clients of the cottages loved calling us. This was the first time Captain Maynard or the Sloth had been there (as far as I knew). We walked up and were greeted by a couple of very thin artists. One of them commented that he loved my sunglasses. They were holding up an older Mexican gentleman who was wearing a cowboy outfit. He was obviously very drunk. The artists explained that they were on their way to their room to play hide the Easter egg when they saw this man wandering around in the street. Being gentle humanitarians, the men put their recreational activities on hold and escorted the inebriated fellow to the safety of the sidewalk. Our tipsy cowboy was very fortunate that he picked this spot to play in the traffic. The patrons of most establishments on this strip of road would be placing bets on what kind of vehicle would hit him—and whether or not he would live—then they would fight over his hat. The good neighbors had originally called the cops because the man refused to stay out of the street; they would escort him to the sidewalk only to have him wander back into the roadway. The police told them they were too busy to come out for a drunk and forwarded the call to us. Captain Maynard told the cute couple that we would take it from here and radioed for the LARC van. (LARC stands for Local Alcoholics Rehabilitation Center. They pick up the winos and take them back to their facility to dry out.)

As the pair turned back toward their room, the Sloth mumbled some derogatory comment about alternative sexual preferences. One of the men turned around and demanded to know what the Sloth had said. I told him not to take any offense because the Sloth was going through a period in his life that had him questioning his own sexuality. The angry gay man became very consoling and told the Sloth that he would never be happy until he came to grips with the fact he was gay. The Sloth was infuriated. He looked at me and said, "I'm sick of your shit. Keep your insipid little comments to yourself. You're a joke. And why don't you ever take off those stupid glasses?" My resilience was paying big dividends. I could picture the increased acid production inside the Sloth's belly. My new gay friends were appalled by the Sloth's comments. The boot held on to our singing cowboy, who was really belting out the Mexican show tunes. Captain Maynard didn't know what to do. I smiled at the Sloth and told him that if he was really a man, he would blow me right there on Baseline Road and profess his new found sexuality to the world. I was secretly hoping he would take a punch at me. One of the gay guys said, "Tonight marks the end of lent, do it!" The Sloth stormed back to the rig. Before Captain Maynard could berate me, LARC arrived and took custody of our drunken balladeer. I thanked the gay couple for keeping Roy Rogers out of harm's way, gave them a handful of latex gloves and told them to have a safe evening. When we got back in quarters, Captain Maynard told me I was forbidden to speak on calls, and I had to clear any actions with him first.

Two shifts later, we had a nasty heart attack patient late at night. We arrived to find an old man drenched in sweat and very pale. He was pleading with us not to let him die. The Sloth told him we wouldn't let that happen, he had the best two medics on the fire department working on him. His tone was saturated with arrogance. This seemed to calm the patient down, which was good. It would have been the wrong place to make a comment anyway, and it was against our captain's policy for me to do any public speaking. The Sloth and Captain Maynard were speaking the paramedic language: "myocardial tamponade, transcending endfarctist, epi, stat, blah, blah, blah." They both loved this part of their job. The boot and I took vital signs, put the patient on oxygen, and set up the IV for the medics. Captain Maynard was bent over the heart monitor when the patient

72

removed his oxygen mask and unloaded his spaghetti dinner. The Sloth wasn't so lucky. He had just finished pushing some type of cardiac drug into the patient's IV and ended up with partially digested pasta from his right shoulder to his shoe. It took every ounce of my resilience not to laugh. Cardiac patients will oftentimes code (die) when they have an emisisotic episode (paramedic lingo for vomit). Our patient told us he felt much better before wiping his mouth and replacing his oxygen mask.

The ambulance arrived, and we helped the crew load the patient into the back. The Sloth and the booter were going to ride along and monitor the patient. Captain Maynard and I were going to follow them to the hospital, which was about 7 miles across town. I jumped in the driver's seat while Captain Maynard rode in the passenger-side officer throne. Our regular rig was in the shop for maintenance. We were in an old reserve rig, which had a manual transmission. All of our frontline apparatus had automatic transmissions, so I was not well-versed in driving these older models. Shifting these prehistoric monsters was no easy task. It required the ancient art of "double clutching." You had to synchronize your engine RPMs and road speed with your shifting pattern. When you shifted from one gear to the next, you had to pause at neutral while you let the clutch out. Once in neutral, you slammed the clutch back in and shifted into the next gear. This entire lunatic motion was supposed to be completed in less than a second or you ended up grinding the gears. It was reputed that the older guys, who drove these rigs in their heyday, could shift without using the clutch. I was not one of them.

Once I got in the rig, I looked for the battery switch. This opened the circuit between the starter and the rig's six large batteries. I found the switch and turned it on. A very loud buzzer went off. This obnoxious feature was designed to let a deaf person know the rig was ready to start. I pushed in the clutch and fired up old Nellie. The entire rig looked and felt like it was going to vibrate into pieces. I looked at Captain Maynard and screamed, "I bet those guys from a couple shifts ago would love to have anal sex while riding around in this thing." He looked at me and said, "Shut up and drive."

I pulled behind the departing ambo and ground the gears going from 1st to 2nd. Traffic was light (it was 2 a.m.), and after a couple of lights I was starting to get the hang of shifting. Captain Maynard kept telling me I was doing it wrong. He was really raining on my parade. I had gone through my third yellow traffic signal when Captain Maynard angrily told me not to go through another one. I made sure the next light I went through was yellow. Captain Maynard yelled, "I am giving you a direct order! Do not run another yellow light!" Looking back, I'm sure he was hoping I'd run the next one so he would have something he could write me up for. He pulled his heart medication out of his pocket and was taking a pill out of the bottle when the next light turned yellow. There was not another car in sight. I double clutched into a lower gear and locked up the brakes. We had been going about 40 mph and were 200 feet or so from the intersection. A fire engine makes a very distinctive sound when it skids out of control. The large rear dual tires make a deep, almost guttural howling noise. The truck had gone into a fishtail and was now careening sideways. We came to a sudden, jerking stop just short of the intersection. Clipboards and books launched out of the rack between the me and my captain. The engine stalled, and the obnoxious buzzer went off. Captain Maynard's heart pills had scattered everywhere. I looked over at my disheveled captain and said, "Boy, that was close." He didn't wait until we got back to the station; he added driving to my "not to do" list while we were still in the empty intersection. He also told me to lose the sunglasses.

☠☠☠☠☠☠☠☠

One of the things I really hated about Captain Maynard and the Sloth was their propensity at avoiding manual labor. Firefighting is very hard work. One afternoon we went to a fire in an auto-wrecking yard. Some kids who lived in the projects next door had set some trash on fire in a nearby ditch. It was a warm day, and the fire quickly spread. When we got on the scene, about six of the cars were burning. This represented less than 1 percent of the vehicles in the 5-acre yard. These kinds of fires really suck. Most of the wrecked cars are destined for the scrap yard where they will be crushed into recycled steel. If you put the fires out, you haven't really saved anything. Our other option was to let them burn, which has a downside, too. Thousands of burning cars are hard not to notice. Occupied businesses and homes bordered the auto-wrecking facility. Our only option was to put out the fire as quickly as possible. We hooked up to a hydrant, and the boot pulled an attack line while I laddered the chain-link fence separating us from the wrecking yard. Once the boot got over the fence and got his line into action, I pulled a second line and joined him. By now, 10 cars were burning. The boot took one end of the burning line of wrecks while I took the other. It took us the better part of half an hour to knock down the main body of the fire. We spent the next 30 minutes mucking. Some of the vehicles still had tires on them, which are next to impossible to extinguish. To make sure the fire is completely extinguished, you have to apply water to every surface. This means you have to get inside the vehicles to tear apart the upholstery, and under the hood to extinguish burning engine components. You have to find out where all the smoke is coming from and apply water until the smoke stops, or you'll be back later to repeat the whole process.

The fire had driven all sorts of vermin from their hiding places. Lizards, scorpions, large spiders and an array of flying insects scurried away from death by fire. The boot and I were watching a half-burned snake slither off when I commented that all we were missing was a goat and a bridge troll. He giggled as he pointed over toward the truck and said, "No we're not." I looked over and saw Captain Maynard and the Sloth sitting under a shade tree next to our water cooler. The Sloth was screaming at some of the neighborhood ghetto children. The kids were running off and letting the two social scientists know they were number 1 by raising the middle fingers on their small hands. Once the impetuous youngsters got a safe distance away, they threw rocks at the dynamic duo. It was my most joyful moment of the shift.

Life for the next few months followed the same course. Captain Maynard and the Sloth tried to kill any joy, and I tried to kill them. I thought I could accomplish this with my cooking. Captain Maynard took his heart medication and lots of antacid tablets. The Sloth was always on the verge of developing an ulcer. I put jalapeños and big scoops of mayonnaise in everything I prepared. It was my way of being resilient. I have to give them credit; they ate everything. One night, after a very spicy meal of red chili burros, we responded to car accident on the freeway that changed everything.

It was after 1 a.m. when the call came in. I rushed to get my turnouts on when I saw a column of smoke and fire rising into the night sky. Just before we got to the scene, we drove past several highway patrol officers who were frantically waving us in. They had gone so far as to close the freeway. I took this as a very bad sign. These guys never closed the freeway. I had been to a call for a burning gasoline tanker a few years earlier where the highway cops had tried to route traffic around the conflagration. Our battalion chief almost came to blows with one of their sergeants before we could get these law-enforcement brain surgeons to close the roadway. They were very zealous about keeping the traffic moving on their ribbons of concrete.

The accident scene spread over several hundred feet. It looked like World War II. Four unidentifiable vehicles were torn to pieces. Motors and transmissions littered the road. The bed of a pickup truck had been thrown about 75 feet up an embankment. It was wedged between a tree and the overpass. One of the vehicles was well involved in fire. Two highway patrolmen were trying to extinguish the burning wreck with fire extinguishers. The fire was so hot the cops couldn't get very close to it. They were in a low crouch, trying to escape the radiant heat. The small, powerful streams of powder coming from the extinguisher nozzles disappeared into the burning mass. When the powder hit the fire, it turned into white clouds that glowed under the sodium streetlights. It shrouded the entire accident scene in a fog that seemed right at home among all the death and chaos. The powder knocked most of the fire down by the time both of the units were completely discharged. Within seconds, the fire returned to its original size.

The Sloth stopped our rig about 25 feet short of the burning car. I stepped off the truck and pulled a 1½" attack line. As I was flaking out my line, I noticed four twisted, mangled bodies in the road. They looked like they had been fired out of a giant slingshot at 300 mph before coming to a sudden stop into the overpass. Some had their heads facing in the wrong direction, while others had their limbs arranged at impossible angles. One of the dead looked like she was doing a final cheer. Her leg was lying on top of her torso with her foot dangling over her head. Her other leg was shooting out from her hip at a 90-degree angle. All of them had bones poking out of their skin. None of them was moving.

It only took about a minute to knock down the majority of the fire in the car. The car's two occupants were quite dead. They were both burned to the extent that most of their skin had been sizzled off. They looked like they had been dipped in tar. The driver had a death grip on the steering wheel at the 10 and 2 o'clock positions. He had died with correct hand placement. The flesh had been burned off his face and he looked like he was smiling. His burned out eye sockets were full of extinguisher powder. Both he and his passenger were still smoldering. As I dropped the line, I went back toward our rig to see if I could help treat any survivors. I looked over and saw Captain Maynard kneeling over one of the obviously dead victims, feeling the carotid artery in his neck for a pulse. It was grossly absurd and comical at the same time. The body was on its back but the face was buried in the road. I'm not a medic, but I would have bet that the guy's heart had stopped. (I still wonder what my new medic captain would have done if the guy did have a pulse.) The Sloth and the boot had gone to the only vehicle with any remaining survivors.

We had been on the scene for no more than 3 minutes. The call had been dispatched with two engines, one ladder and a battalion chief, but we were the only fire department unit on the scene. A man and woman were trapped in the final wrecked vehicle. We would need the extrication services of the ladder company to get them out of the heavily damaged car.

We had six dead and two critically injured trapped people. It had taken us less than 2 minutes to get to the scene from the time we were dispatched. But there was far too much work for a single company to do. In most other lines of work, this problem is addressed by getting a cup of coffee and waiting for more help, calling the boss, or putting together a focus group. For six of our evening's customers, it didn't really matter. They were done with time and free of its constraints. The other two patients had very little available time, and if we didn't intervene on their behalf very quickly, they would soon be joining the six others who hadn't survived the nasty encounter.

The need to perform miracles in seconds has a tendency to put a bit of pressure on the incident players. If you get sucked into this time-compressed vortex, you can easily get swept away in its currents. You become so overwhelmed with all the things that need to be done that you become ineffective. It's like the man said, "Once you lose your head, the next thing to go is your ass." When most of the incident customers are taking what could be their final breaths, paramedics may feel a little more pressure than their Neanderthal BLS brethren do. Medics have been entrusted with the tools, potions and secrets of Hypocrites. The good medics focus on one or two things at a time. If the patient is having trouble breathing, they start with that and try to make it better. Everything beyond the breathing is just noise and they shut it out. Screaming, moaning, burning flesh and headless patients are not distractions. The only thing in their world at that moment is the breathing. Once the breathing is fixed, they move on to the next most likely thing that will kill the customer. They are very clinical. They do not get excited or yell. Some people mistake this for ambivalence. If you didn't know any better, you would think they were performing some mundane task, like painting a house, not saving someone's life. They are really good at what they do.

The Sloth was barking orders at the booter who was trying to hold a trapped patient's head straight and keep his airway open. Both of these things are pretty important medical interventions for sustained breathing and avoiding any further spinal injury. I don't remember the specifics of the Sloth's diatribe, only that it didn't make much sense given the circumstances. The Sloth then yelled at Captain Maynard to come over and help us with the viable patients. This must have upset Captain Maynard, because he came over and ordered the Sloth to move the truck for some inexplicable reason. The shades were coming down. The booter gave me one of those "Do these guys always lose their minds?" looks. Ah, the innocence of youth.

I was telling Captain Maynard that both patients were Level 1s and the two in the other car had burned to death when the ladder company showed up. They made quick work of cutting the roof off the car and removing its doors with their powerful extrication tools. Once we had access to the patients, Captain Maynard and the Sloth (having ignored his captain's order) began ALS treatment on the most critical one. They both seemed to have regained their composure. The booter and I went to the other patient whose legs were wadded under the vehicle's dash. The ladder company used their cutters to slice through the car's A-post and rocker panel. Then they placed their telescoping ram under the twisted dashboard and against the bottom rear corner of the doorjamb. They spread (or as we say, rolled) the dash off the patient. It had taken the auto-shredding clinicians less than 10 minutes to remove the two occupants from the badly mangled car. By the time we got the patient out of the vehicle and onto a backboard, our second ALS company had arrived. The patients had what the medics like to call "multiple systems trauma." They had broken bones, internal bleeding and damaged organs. A couple of months later both patients would end up walking out of the hospital.

The next morning when I woke up, Captain Maynard was waiting for me. Every fire station gets lights and a wake up tone at 7 a.m. Many of us ignore this tone and simply roll over and go back to sleep. This tone has outlived its usefulness, but ancient customs die hard. Captain Maynard wasn't known for being a late sleeper, so it surprised me when he was still there at 8:30. As I poured a cup of coffee, he asked if he could talk to me about something. We were the only two in the station because C Shift had already left to play racquetball for their morning physical training. We sat at the kitchen table. He thanked me for all I had done at last night's call. Then he went off on the Sloth. He was not happy with the way the Sloth performed under duress. I had a pounding headache from breathing in the smoke from the burning cars and people, so I just sat and nodded, reminding myself that SCBAs were good things and not to be so stupid as not to wear one next time. I wondered to myself if one underwent some type of profound karmic change after breathing in the incinerated essence of other people. Captain Maynard gave me back all my privileges with the exception of driving. I told him I appreciated his newfound confidence in me, and I told him if he would quit screwing with me, I would get rid of the sunglasses.

The next few months around the station were not as tense, but I still

78

wouldn't describe them as joyous. Captain Maynard pretty much left me alone, but he continued to be a miserable person to live with. He had also taken the Sloth into his office for some private one-on-ones. This seemed to fuel the Sloth's hostility. He couldn't take it out on Captain Maynard, and he no longer enjoyed parrying with me, so he lashed out at the booter whenever the opportunity presented itself. I still enjoyed coming to work, but I was looking for greener pastures. I had worked with these two long enough and felt I could leave with my resilience intact. A couple of my associates were fighting over a vacant position on another engine company. I had more seniority than both of them and had decided to put in for the spot. I told Captain Maynard my plan. He felt it was a marvelous idea and thought that I would fit in nicely with the other criminal types assigned to that particular station. He was almost finished with his tearful goodbye speech when the lights came on and the voices told us that the dump was on fire. That was one of the problems with our first-due area. A lot of the things that burned were not worth saving.

We went out, got on the rig, pulled out of the station, hung a right and headed for the dump south of the station. There was a mulch-processing plant across the street from the dump. The southern border of these two tourist attractions was the Salt Riverbed. It had been dry since they built the Roosevelt Dam and created Lake Roosevelt about 30 years earlier. The riverbed was home to rock quarries, concrete businesses, homeless people, the random dead body and an occasional pile of illegally dumped trash. We avoided going down there at night because packs of wild dogs had claimed it as their territory. We would routinely get fires in the river bottom. If they were small enough and didn't threaten life or property, we'd let them burn.

As we made our turn out of the station, I could see a large column of smoke toward the front of the dump. Usually when the dump burned, the smoke was more whitish in color and not very dense; it also tended to spread over a large area. This column of smoke was tight, dense and black. It looked like smoke from a structure fire. My steel-trap mind quickly determined it was either a vehicle fire or one of the dump's offices was burning.

Our rig came to a stop at the entrance. The boot jumped off the truck to wrap the 4" supply line around the hydrant and make the hookups. The rest of us proceeded to the dump offices, which were housed in a large doublewide trailer. Fire blowtorched from all the doors and windows. I got off the rig, opened the SCBA compartment, grabbed my mask, threw it on and donned my face piece. The engineer (not the Sloth, who had taken the day off and was probably at home killing bunny rabbits or giving massages to Cub Scouts) put the hose clamp on the supply line. Captain Maynard was donning his mask. We had had more than 25 structure fires since Captain Maynard came to Engine 55, and this would be the first time that the two of us went into in a burning building together. I was going to have some fun. I pulled a 1½" attack line and waited for Captain Maynard. He came up behind me, and we headed toward the burning door. I wish I had given him the line. That way I could have used him as a heat shield, and he wouldn't be in a position that offered him retreat.

I was 5 feet from the door when I opened the line. A wide fog stream came out of the nozzle. I adjusted the nozzle to a straight stream and played it into the burning door, working the nozzle in large clockwise circles. The fire darkened down. The water quickly converted to steam and knocked some of the energy out of the fire. When water is applied to an out-of-control fire, it turns into steam. Part of this transformation affects the amount of space the steam can fill in relation to its previous liquid form. One cubic foot of water can expand into more than 12,000 cubic feet of steam (give or take a couple thousand cubic feet for all you science fans). This steam conversion works especially well when the steam cannot easily escape the burning area. If the fire is contained in an area with six intact sides—usually four walls, a floor and a roof—the steam smothers the fire. I have been witness to many well-involved room fires that were knocked down with less than 10 gallons of water.

The fire in the doorway was replaced with dark black smoke and big, white puffy clouds of steam. I continued to apply a straight stream of water toward the ceiling and advanced the hoseline inside the trailer. Captain Maynard hung with me for the first few steps. It was quite toasty inside, hotter than it had been outside, when my turnouts first started smoking. I continued to apply the powerful stream of water, and then a funny thing happened at the office. Captain Maynard bailed on me just as the large room I had just extinguished burst into pretty orange flames. It felt like wasps were attacking my shoulders. I dropped to the floor and blasted the ceiling with water. The pretty orange turned black again, then back to billowy, white-hot steam. As I lay on my back, I continued with the onslaught of water.

When the fire flashed over, I was at least 10 feet inside the trailer. My plan had been to join Captain Maynard on the outside, but my self-preservation efforts prior to exiting had extinguished the fire. The fire had flashed on us because the moron who set it had used 10 gallons of gasoline. He had been fired that morning from his job at the dump. (I wonder what you have to do to get fired from the dump.) His bad day had just begun, however. He

80

went over his girlfriend's house to find some solace, and when he gave her the bad news, she showed him the door. You could say she dumped him after the dump dumped him. Sick of the entire dumping process, he returned to the scene of his frustration to get a little payback. His day would get a lot worse when the police charged him with arson in an occupied structure. This is a very serious felony charge with a mandatory prison sentence. At least he would be going to a place where one didn't get to dump their boyfriend when times got tough.

When I came out of the now-extinguished office, I noticed my turnout gear was smoldering and the whole world looked rather distorted. I took off my helmet. The face shield had melted and dripped onto the lens of my mask. I pulled back my Nomex hood and removed my face piece. My SCBA mask was still flowing air, so I turned off my regulator and slipped the SCBA off my back, throwing it on the rig's bumper. When I took my turnout jacket off and dropped it to the ground, I could see it was heavily scorched and had a large hole burned into its left shoulder.

Turnout gear has undergone huge advancements during the course of my career. The ancient elders who raised me started their careers with cotton-duck turnout jackets, Frisco jeans, fiberglass helmets and cotton work gloves. This ensemble offered marginal protection. The fire in the dump office would have killed a firefighter 30 years ago. Today's protective envelope offers 20 seconds of protection from a 1,200-degree flashover without serious injury. Although today's protective gear is much better, it still has its limits. It will protect you from very high temperatures for brief periods of time. At lower temperatures, it will protect you longer but not forever. I had just gone through a brief flashover. My gear had been exposed to several minutes of temperatures above 400 degrees and around 10 seconds of 1,000 degrees.

My epidermis paid the price for my stupidity. It is not unusual to come out of a hot interior firefight and be all pinked up. It's a lot like having a sunburn and usually fades after several hours. It also isn't out of the ordinary to feel a little tingly all over after wearing a flashover. After I took my jacket off, I closed my eyes and spread my arms to let the 100-degree breeze wash the heat off of me. It felt like I was standing in front of an open freezer. I looked down at my blackened, fiberglass-wrapped SCBA bottle and could see heat still radiating off of it. The sides of my face stung and my left shoulder started to throb. The boot came over and looked at me. He said I had a few blisters on my face. I pulled the left sleeve of my tee shirt up. It looked like someone had cut a large grapefruit in half and attached it to my shoulder.

My plan had been to take Captain Maynard on a stroll through a world that he scorned. I didn't want to maim or injure him, just remind him that he worked for a FIRE department that also went on EMS calls, not an EMS department that showed up to fires so they could watch other people work. It would also give him something new to bitch about. Instead of teaching my captain a lesson, which in retrospect was an idiotic idea to begin with, I had ruined any chances I ever had at being a male model.

I got a free ride to the hospital burn unit. They dressed my burns and sent me home to recuperate and heal for the next three weeks. About a month after I went back to work, I got a letter from the State Compensation Fund. They requested my presence at their offices so they could close out the file on my little dump-party fiasco. I met a very concerned woman at their office. She wanted to make sure I had fully recovered. We were having a pleasant chat when she noticed some scars on the sides of my face. The steam had blistered thin lines from my temples to my jaw. The burns were pretty well healed. Only a little discolored skin remained. She informed me I had "permanent facial disfigurement." I replied that I was born this way; she would have to talk to my parents. She giggled and told me she would take care of everything. A month later I got a check in the mail. Leave it to the government to compensate someone for an act of stupidity.

That was my last call on Engine 55. I learned one needs to temper their resilience with good decision- making skills. It was brainless to incinerate my body at the dump. I never received another serious burn after my misguided kamikaze mission. An associate of mine took my spot on Engine 55 after I moved on to the greener pastures of a station in downtown Phoenix. The night before he went to 55, I met him at a local watering hole and gave him a little gift—my old sunglasses. He wore them to work his first shift. Captain Maynard's nightmare wasn't over; it just wore a different face. Captain Maynard only lasted a couple more months before he found a nice slow spot in the suburbs, where he finished out his career. The Sloth followed him a few months later. Captain Maynard retired years ago. He didn't get all of the contributions he made into the pension system before he had a life-ending heart attack. I haven't seen or heard from the Sloth in 20 years. His whereabouts are currently unknown.

Chapter 7
Tacos & Carnivores

Station 66 is located east of downtown, about 2 miles north of Sky Harbor Airport. The station itself was built in the early 1970s and was constructed like a tract home with a really big three-bay garage. Some described our firehouse as "intimate"; others used phrases like "falling apart," "dump" and "piece of shit." The place really fit into the neighborhood. It was always infested with some type of vermin. First it was roaches. A few weeks after the roaches were eradicated, our dishwasher broke. The service guy showed up with a replacement unit and found a nest of coral snakes living under the old one. During lunch one day, hundreds of tiny spiders descended from the holes in the acoustic ceiling tiles. Next came the mice. After one late-night call, I was walking into my office in a fruitless attempt to get a few hours sleep. When I turned on the light, a blur shot across a cabinet and jumped to the floor, knocking over a 5-foot-tall fake tree in the process. It was not a mouse, and I don't think it was a cat. I closed the door and spent the night in a recliner.

Station 66 is sandwiched between the County Hospital and a potter's field owned and operated by the state hospital for the criminally insane. Our first-due area was ground zero for a variety of halfway houses, drunk tanks and detox centers, and, for a short period of time, the Arizona branch of the Republican National Committee. The locals didn't need to leave our first-due area to secure their drug of choice, overdose on it, have the medics from Station 66 bring them back from the dead, and be transported to the county hospital before being transferred to one of the many neighborhood clinics or halfway houses. The addicted accounted for a good portion of our station's customer base, but they were severely outnumbered by the crazy.

The crews of Station 66 had the immense fortune of delivering service to one of the lowest economic rungs of society. Our first-due area was quite literally a freak show of the human experience. At night, the streets came alive with prostitutes, transvestites, gang-bangers, drug dealers and the collaborators who follow this type of nocturnal carnival. It was a great place to work. We had shootings, stabbings, substance-induced car accidents, and fires — lots of fires. We were mercenaries wallowing in the chaos of our 24-hour shift; if you got three hours of sleep in a row, you were lucky.

During this "good old days" era of my career, I worked on Ladder 66 with two very talented engineers, Jake and Luke. Luke possessed the patience of a saint, while Jake had a natural connection with the insane. A never-ending string of probationary firefighters rounded out the fourth spot on our ladder company. Each of the new kids spent a three-month stint with us. It was better than going off to camp for the summer.

The personality of B Shift's engine company was fueled by its captain and engineer. Captain Sean and Engineer John (my brother) had the bodies, dispositions and hairstyles of Marines. What the two lacked in patience, they more than made up for with a glut of competence. The alpha-male pair in the front seats of Engine 66 balanced nicely with the two polite medic firefighters who rode in the back of the rig. Patrick and Jesse were as easygoing as Sean and John were intense. The group of them made a lovely quorum.

B Shift maintained a healthy and caring relationship with its customers. In the event the incident participants were medically stable and mentally crazy, our crew would mirror those behaviors. It's how we maintained our sanity. It also made the incident fun for everyone. I have never been fond of members of the public-safety community who minimize the needs of customers who call for anything less than a life-threatening emergency. These same public-safety professionals also have a tendency to disregard bystanders or other interested parties who hover on the periphery of our incidents. It's simply rude to shoo off some Good Samaritan who just wants to help or provide a piece of useless information they believe will make the difference between life and death. Really, if there isn't an imminent hazard or some pressing action you must take, why not slow things down and donate a couple of seconds to these folks? It doesn't cost anything, and it will make them feel better about themselves. Besides, you're on the call anyway. What good does it do to get pissed off because someone isn't bleeding to death or a building full of cooing baby orphans hasn't burst into flames? Nothing will please a lunatic who needlessly calls 911 quite like a kid's red plastic fire helmet and a pocketful of brightly colored fire-hydrant pencil erasers.

I have worked calls both ways. I'm not proud to admit I've lost my patience and gone into A Shift-style rants, lecturing the patient about wasting our time with their petty whining and snivels. This does no good. Does the customer care that our lights-and-siren response to their "problem" put the general public at risk? Do they care that while we are tied up on their needless call we can't respond to and the save the life of a pregnant woman who was just hit by a bus? Do they care that we are smart and they are dumb? The answer is no. Crazy customers who call us because their alarm clock won't keep the correct time, or because giant cockroaches are stealing their bike couldn't give a shit about our lectures. The solutions to these problems are simple: Don't put the public at risk by driving like a madman when responding to any type of call, monitor the radio for emergencies while you are taking care of non-emergency business, and never forget that none of us is as smart as we think we are.

I discovered the practice of voodoo medicine is an excellent way to deliver high-quality service to the mentally unstable. I kept a small sack loaded with dice, chicken bones, dried flowers, ammonia inhalants and a vial of Tic-Tacs next to our EMS clipboard. In 10 minutes of total on-scene time, my crew and I could diagnose an illness, remove spells, tell fortunes and dispense minty-fresh medicine. The ammonia inhalants brought balance to our rituals, sending a veiled message to the non-emergency patient that we could always resort to the powerful antidotes of modern medicine if need be. We performed these rituals of spiritual magic several times a shift for a couple of months until C Shift threw our bag of charms away. Each occurrence of Santeria-style ministering resulted in a happy customer and a happy ladder crew. This is the highest level of satisfaction ever recorded for any type of government work.

After the C-shifters disposed of my bag of voodoo trinkets, we B-shifters had to improvise. One of our recurring customers was an older gentleman who had a forest of black hair growing from his nostrils. He spent large hunks of his waking day rummaging through alleys in search of discarded writing devices. He never went anywhere without his collection of hundreds of pens and pencils. I had dozens of encounters with this man and never once knew him to be sick or injured or to actually write anything with his coveted relics. Our weekly customer service get-togethers with the Pen Man usually involved throwing the bones and dice to determine the location of a hidden mother load of unused Bic pens. Sometimes the Pen Man would stop by the station for a reading; other times he would approach us while we were on another call. Occasionally, a passing motorist would call because they couldn't determine from their speeding car if the Pen Man was lying dead on the sidewalk or simply sleeping. We rolled on the Pen Man dozens of times. All he ever needed was a 10-minute reunion. The voodoo just made it fun for everyone. It doesn't take a herd of high-priced consultants to figure out that all these crazy individuals want is some quality human interaction.

It has been said that dog owners grow to resemble their pets. This same phenomenon holds true for firefighters and their first-due areas. The plastic veneer that holds polite society together had long ago been peeled away from our little hunk of the city. Our station responded on over 10,000 runs annually with three pieces of apparatus. The unique clientele combined with the high call volume amplified any of our personality quirks along with our most deeply held beliefs. The religious turned more deeply to God in an effort to make sense of the spectacle that surrounded them. The laid-back fought a daily battle against full-blown apathy. Overachievers pressed their uniform T-shirts and studied endlessly for promotional exams.

The most powerful coping element firefighters use to stave off an insane world (and first-due area) is the bonding process. Fire stations more closely resemble a home than your traditional workplace. Many social scientists attribute society's impending cultural demise to a lack of quality family time. The hectic pace of modern life is slowly killing family rituals. The fire service is one of the last bastions of the home-cooked family meal.

One of the benefits of a 24-hour shift is that it allows us to eat lunch and dinner together. Most employers view lunch as a federally mandated loss of productivity during the middle of the workday. Truth be known, shopping, cooking and eating together is a more powerful group bonding experience than a million dollars worth of moronic corporate team-building sessions. In our business, firehouses that don't eat together are viewed as severely dysfunctional.

I did not look forward to the cooking task, but many times the alternative was much worse. I am by no means a chef, but I can prepare edible food. I have found it is better to leave the sacred ritual of preparing our daily meals in competent hands as opposed to entrusting it to someone with question-able abilities.

Cooking two meals for 10 firefighters is an all-day activity. You have to decide what to make, collect chow money from everyone and do the shop-ping. On my day to cook for Station 66, we prepared a lovely lunch of grilled-cheese sandwiches with green chilies and homemade clam chowder. After we cleaned up the lunch dishes, we started on the evening meal: corned-beef tacos. This gastronomical delight required no side dishes. The mere mention of corned-beef tacos makes strong men leak urine. They are rumored to bring redemption to the damned.

We make a dozen tacos per man; 120 tacos are a five-hour labor of love. We start by baking 7 lbs. of corned beef for 4 hours. While the meat cooks, one of us fries the corn tortillas into the folded taco-boat shapes. This tedious and hazardous chore takes 90 minutes and causes at least one second-degree burn, but using store-bought taco shells is unconscionable.

Next, we dice 5 lbs. of potatoes. We fry these (naturally) with onions, red bell peppers and a mix of chilies. After seasoning the potato mixture, we set it aside to cool. Then we make the hot sauce. We blend tomatoes, green onions, jalapeños and Serrano chilies with a multitude of spices, including a good dose of cumin. The coup de gras is a couple of habañero peppers. After the meat is done, we shred it and thoroughly incorporate the potato mixture. We place 3 tablespoons of this hash mixture into the taco shells, and then we sprinkle the tasty payload with white cheddar cheese. We bake the crispy corn shells for about 20 minutes, and then top them with diced tomatoes and a mixture of shaved iceberg lettuce and cabbage dressed with fresh lime juice and white vinegar. This boat of love is topped with the sweet and tasty hot sauce.

Taco night always turns into a Roman feeding frenzy. At the peak of consumption, the diners are transported to a different level of consciousness. The only way we could get the tacos down any faster is if a small boy shot them down our gullets with a wrist rocket. We always try to eat early so we can squeeze in a pre-bedtime feeding before going off to dreamland.

Eating gourmet meals, working out to your heart's content and tribal socialization can cloud a guy's thinking, causing him to believe he belongs to a clan of warlords. Despite these distractions, our core mission is to go on calls and deliver service to the communities we get paid to protect. While the social misfits in our first-due area were usually entertaining, they seldom rose to the level of fascinating. None of the human oddities in our first-due area were quite as strange as the Lion Lady. I met her for the first time at 3 a.m. on the night I made corned-beef tacos for dinner.

Gorging on the delicious feast had lulled me into a deep REM state. My dream world was interrupted when the lights came on and the voices told me we were going on a car accident. As we finished up the call and drove back to the station, we were picked up for an ill diabetic. The ambo was just going available from the hospital, so they also got added to the dispatch. Engine 66 was going back in service from a hospital a few miles away, and they got included as the closest ALS unit. Ten highly trained emergency professionals against one ill diabetic: Diabetes didn't stand a chance.

The ambulance got to the scene first. We parked on the street about 20 feet behind them so the crew would have room to load the patient into the back of the ambo if necessary. I remember seeing my breath as I stepped off the rig with my clipboard and portable radio. The temperature was in the mid-40s, but dinner's hot sauce still warmed my gut. It was a moonless night, and the streetlight in front of the house was burned out. The only light on the street came from our apparatus. The ambo's back doors were open, and the gurney was gone. I got past the front of our rig and turned to walk up to the house when I noticed the gurney lying on its side on the walkway leading up to the front door.

The front door stood wide open, there were no lights on inside. The place appeared abandoned. The rest of the ladder crew was getting the EMS stuff off the rig when I made my way to the front door. I poked my head into the dark living room and called out for Mike and Biscuit Head, our 500-lb. ambo crew. They each hovered around 6'4", were very strong and rarely spoke or smiled. I took a step into the house and stopped to let my eyes adjust. It was darker inside than outside. I kept calling "Biscuit Head...Mike..." and got no reply. The front window of the house had a heavy drape over it. The only light that made its way into the house was the steady beat of red and yellow made by the rotating lights on our ladder truck. The living room contained the usual assortment of broken-down furniture. I found the light switch by the front door and flipped it. Nothing. No light, no patient, no Mike or Biscuit Head.

I was midway into the living room when I asked, "Is anyone here?" To the right was a hallway that led to the bedrooms. Straight ahead were the kitchen and dining room. I turned to the right and looked down the dark hole that led to the sleeping chambers and decided to try my luck at finding human contact in the kitchen. There was a window in the kitchen door, which led to a covered backyard patio. The rear of the house bordered the back of a large grocery store. A large light hanging on the back wall of the store introduced a small amount of sterile white light into the kitchen area. I had made it to the end of the living room and was standing by the small table in the dining room when I noticed movement in the shadows of the kitchen.

I stopped and said "fire department." The shape was drawn to the melodic ring of my voice. It made animal noises and started to shuffle toward me. I started backing away. It was still hidden in the dark crater of the kitchen. I was now standing in the middle of the living room. The dark shadow dragged one of its legs as it continued to advance toward me. I thought for a moment that this thing had eaten Mike and Biscuit Head. I grabbed my $2,500 Motorola portable radio by the antenna, planning on using it to club the dark mass advancing toward me. This thing was not going to take me without a fight. I screamed at the lumbering savage to stop. It was standing next to the dining-room table. The beast moved into the light filtering through the kitchen door. It had an enormous head that capped its twisted, hunchbacked frame. Its arms hung to its knees. I could see light shining through its nightgown. Our red-and-yellow emergency lights intermittently washed across the apparition. Just when I thought I could make out its features, darkness replaced the yellow lights, and then the night would fill with half a second of red. This constant cycle of yellow, black, red would have had a hypnotic effect had I not been terrified.

I was standing in the doorway screaming at the shape, telling it to not move, when the rest of the ladder crew joined me. Jake and Luke had retrieved the abandoned gurney in the front yard and walked in with our booter. When they asked me what was going on, I simply pointed at the scary ogre. Biscuit Head and Mike finally appeared. I asked them what they were doing. They both looked like little girls who had just seen their first snake. "Did you see it?" Mike asked me. I nodded toward the open door. They approached the door as if they were taped together and had rats in their asses. As the six of us stood gawking in the doorway, Luke shined a flashlight on the home's occupant. She was very deformed. Her head had the general shape and dimensions of a large jack o' lantern. Her big eyes were offset at odd angles from one another and covered with white cataracts. Her nose was basically two holes above her crooked mouth. Sparse, stringy, matted black hair barely covered her misshapen skull. She spastically jerked her head back and forth like a bird looking for a big, fat worm. In a strong whisper Biscuit Head said, "Not her. It's the other one. You won't believe it. The other one in there ain't normal." I looked at Biscuit Head and said, "You got to be shitting me. You're calling her the normal one?"

I returned to the person standing in front of us. I walked in and asked her to sit down. She did as I asked. She missed the chair by about 2 feet and ended up bouncing off the floor. Jake, Luke and our booter went over and sat our patient up and started the treatment process. She was blind (which was the reason she missed the chair) and her blood sugar was low, which put her in a depressed mental state and caused her to babble. Later on, Jake would say, "Well, she wasn't much to look at, but all little sugar booger needed was some orange juice, and all you big strong firemen could do was run around like a bunch of 'fraidy cats." I started to fill out an EMS-encounter form as Biscuit Head and Mike wheeled their gurney into the house. They both looked very tentative.

I noticed movement out of the corner of my eye coming down the dark hallway. Biscuit Head and Mike both screamed in unison, "It's her! It's the Lion Lady!" as they fled from the house for the second time. I turned and came face to face with the mother of our deformed patient. It was obvious our patient suffered from an array of birth defects. Her mother, on the other hand, appeared to have been created by a mad scientist. I would not begin to venture a guess at how old she was. She had too much other stuff going on to care about her age. She was a wonder. She stood 5'10" and her clothed body appeared completely normal. The woman had long, bushy, flaming-red hair. She didn't have a mouth as much as she had a snout. Her mouth was pulled out a good inch beyond the tip of her nose. It looked like someone had bred a lion with a human. She also had an unusual amount of facial hair. She was by no means a bearded lady, but she had managed to scare our Hell's Angels ambo crew. The one feature that brought the whole package together was her set of very large eye teeth. I would not have been surprised if she had had been snacking on a freshly killed zebra somewhere in the back of the house.

The daughter had shocked me. Part of this had to do with the whole haunted-house setting. The mother, on the other hand, was very intriguing. I sat down on the couch with her after she turned on some table lamps. The ladder crew put our ill diabetic on oxygen, took her vitals and helped her drink a glass of OJ. Half the juice ran down the front of her sheer night-gown. It was quite a sight. The ambo crew stood with their backs against the wall, terrified by the entire scene. Biscuit Head blurted out, "Are there any more of you?" Lion Lady glared at him. I patted her on the hand and told her not to pay any attention to the two young upstarts. "They are young and not yet wise to some of the special gifts the world has to offer." She flashed me a very toothy smile and almost seemed to purr. Jake rolled his eyes.

The daughter was starting to speak in coherent sentences, so we cancelled the engine and their dazzling paramedics. Lion Lady told us this was not an unusual event for her daughter, and she did not want us to take her to the hospital. Before she finished this statement, Mike and Biscuit Head were out the door with their gurney. We gathered our things, wished mom and daughter well, and went on our way.

It was 4 a.m. when we got back to the station. Jake immediately found Mike and Biscuit Head so he could tease them about hiding from a couple of women. As I walked into the kitchen, I met the engine crew. They had just gotten back in quarters and were pulling out all of the leftover tacos. Within a few minutes, the 10 of us were all sitting around the table, laughing and eating dinner for the second time that shift.

Chapter 8
The Yellow House

I was sitting in my office when the lights came on and the dispatcher announced we were going on a drowning. My partner Dexter and I got in the battalion chief wagon and followed the rest of Station 51 to a home just behind our firehouse. It was quite a contrast. Fire Station 51 had been replaced a few years earlier with a brand new station. The building had state-of-the-art everything, symbolizing and securing its place as an integral piece of the Phoenix Public Safety empire. The exorbitant cost of the new fire station was one of the best investments of public money in Phoenix history. The station is one of the busiest firehouses in the western United States and responds on more drowning calls than any other fire station north of the equator. During the five years I ran out of Station 51, B Shift alone responded on more than 30 drownings.

Generally when we go on drownings, there are quite a few people waving frantically in front of the house. This time there was no one out front. The domicile looked as if it belonged on a nuclear test site somewhere in the Nevada desert. The front yard was landscaped with dirt, which went quite nicely with the bright yellow paint. A few straggly weeds grew into the cyclone fence that guarded the dirt yard.

Engine 51's crew was getting its first-aid gear off the truck and heading to the backyard. I followed everyone into the back, where the dirt motif continued. There were new bicycles, plastic forts and the other accessories one would except to find at the home of two young boys. As I came around the corner, I almost stepped on a small, mangy dog. I avoided the dog, but managed to step squarely on one of many piles of dog shit littering the backyard. Looking back, I have marveled at how that stench seemed to complement the details of the call.

With the repugnant essence of dog shit assaulting my nasal passages, I looked up and saw an old woman in a cheap pantsuit standing over a 2-year-old boy lying on the deck next to the swimming pool. Engine and Rescue 51 immediately started working on the kid. It was evident he was at least clinically dead. The dead always look different. A worn-out man with a gray beard was walking toward the old wooden fence separating his backyard from this cursed home. He stopped at the fence and put his bare foot on the horizontal two-by-four that ran across it, about 4 feet off the ground. He grabbed the top of the fence with both hands and stepped over it into his own yard with one giant stride. He did this as easily as one would get out of a car. Things were shaping up to become a regular carnival. I asked one of the six cops standing in the backyard who the tall-stepper was. They told me he had pulled the kid out of the pool.

The boy's grandmother was explaining to Bill, Engine 51's captain, that she hadn't seen the lad for the past 20 minutes. While 51's crew worked on the dead boy, I went into the fenced pool area to talk to the grandmother and get her out of Engine 51's way.

The water in the pool had evaporated more than a foot below the tile border. Algae had stained the exposed plaster a deep green color. It looked more like a pond than a swimming pool. I couldn't see the bottom through the green soup. Somebody later said it was like looking into smoke.

I looked up as one of the firefighters removed the bag-valve mask because the lifeless patient was puking up the contents of his stomach. I returned to grandma. She was talking to one of the cops about the drowned boy's 4-year-old brother. The police officer told me they had been called to the scene to search for a missing boy. I asked granny if anyone had located the lost brother. She said he was out front. The officer went to the front yard to get the missing lad. He came back a minute later with a 6-year-old kid on a chrome bike. Granny said that wasn't him. The cop yelled to the officers and firefighters at the pool that the 4-year-old was still missing.

One of the phenomena exercising its influence over our work environment is the space-time continuum. It usually manifests by slowing time. These mystical events happen to firefighters whenever the combustion process changes dramatically. Having the room you're crawling through flashover in superheated hues of brilliant orange can stretch your concept of time. These events begin with surprise, evolve into fear and conclude with giddiness once you realize you have survived. The group of us wasn't about to have this kind of time curve.

The time shift at the yellow house began when Engine 44's ALS crew got to the scene. They immediately went to work on the 2-year-old boy. I radioed our alarm room and requested another ALS engine and a rescue. Don, Engine 51's engineer, came around the corner with a 12-foot pike pole. Firefighters and policemen had picked up the long-handled pool brush and a skimming net to frantically dredge the pool. All of these events occurred within a few seconds of one another.

We had been on the scene for a little more than 5 minutes. It was obscene for us to consider another kid had been in this wretched vat of green shit and we had nothing about it. I grabbed the grandmother and asked her if the two boys had been together. She told me, as calm as the tax collector, that she really didn't know. She had been making dinner for the last half-hour and lost track of them. My involvement with her ended when she told me she had found Danny, her 2-year-old grandson who currently had a 200-lb. firefighter doing CPR on his dead body in an effort to make him undead.

It takes most people a while to process the sudden death of a loved one. The fact that this poorly dressed grandmother was completely oblivious to the finality of the situation, and probably suffering her own time-induced disorientation, didn't lessen my nearly uncontrollable urge to throw her into the pool. I looked past the large moles dotting her face and stared into her eyes. She was gone. Her dead grandsons will chase her for the rest of her life. I went back to the pool.

People who have had near-death experiences report reliving the details of their lives. Most of these individuals died for only a few minutes, yet they revisit the decades that were their earthly existence. Time can be a hinky bitch that way. The next 5 minutes I spent poolside felt like a really long week. At that moment, it seemed the only reason the group of us existed was to make a vain attempt to save a 2-year-old boy's life. Even more odd was none of us had even known the kid existed a scant 5 minutes before.

The sight of four large firefighters surrounding a soaking wet dead kid is fundamentally wrong. One of the firefighters was doing chest compressions with the palm of a single hand. He had to be careful to push just hard enough to compress the sternum less than 1 inch into the heart. Any more than that and ribs break, the heart bruises, and lungs can be punctured. While one firefighter manually used the heel of his hand to make the small, lifeless heart beat, his partner squeezed a high concentration of oxygen into the kid's dead lungs with a pediatric bag-valve mask. A medic held the boy's arm in one of his large hands while attempting to thread a small IV needle into an impossibly tiny vein.

The medics were sticking tubes down the 2-year-old's throat while their co-workers kept stirring the muck in the pool, looking for his brother, when it happened. Bill held the pool net, and the long, blue aluminum handle was bowed with some kind of weight on its end. As he raised the net out of the swamp, a small 4-year-old boy appeared. He was face down across the net's plastic frame. He was wearing a white shirt and jeans.

At that moment, I was struck by a vision of this dead kid running around his backyard. His gait was typical of a little boy. It was a choreography of hopping, bouncing and running. Both of his arms were rhythmically pumping up and down, keeping cadence with his short stride. His jeans were tucked into his miniature cowboy boots. For a few seconds, it felt as if the kid would shake off death and go play in his plastic fort.

My earth vision suddenly returned to the backyard and all its evil. The water was still draining off the child's body as Bill swung the pole toward his waiting crew. Pete, the firefighter on Engine 51, grabbed the back of the dead kid's pants and laid him by the side of the green hole that had eaten him and his little brother. Everything seemed very still. I felt as if I were standing in a soundproof booth. Everybody seemed used up. This pause in time didn't last a second, but it hung on everything. It was like an endless moan that never escapes the lips. The mind oftentimes records these events as a series of still pictures. On the other hand, the four news helicopters hovering over the scene were going live with video.

Most drowning incidents happen to young children. While no one looks forward to these calls, after you go on one after another you get used to them. The vast majority of drowning victims we manage to save are forever changed. They live the rest of their lives in a special electric bed and need to have their airway suctioned hourly. The remainder of their lives is spent staring at the same spot on the wall. Very few escape whole.

As I stared at the spectacle unfolding in the backyard, I spent several fractured seconds recalling a story I heard at the beginning of my career. I hadn't thought of this decades-old tale since I heard it almost 20 years earlier, after my first drowning incident. We had just revived a 3-year-old girl, and she ended up suffering no damage from her brush with death. After we finished our call and were back in the sanctuary of our station, Wes, an old engineer, told us about his first drowning. It had happened 20 years prior on Halloween day.

It was around dinnertime when his station got lights for a child in a pool. When they got to the scene, Wes noticed the manicured lawn and the home's neat exterior. As the crew walked up to the front of the house, a frantic clown shot out the front door screaming, "This way, this way!" Another clown toting a little boy's lifeless body followed the first hysterical clown. Both parents had been in the house getting ready for a Halloween party when their young son went out back and fell into the pool. Neither one of the parents was able to give the crew any information; all they could do was cry. The engine crew did CPR on the child in the driveway as the sobbing clowns looked on. After Wes helped load the boy in the ambulance, he turned to head back to the rig. As he spun around, he made eye contact with both parents who were staring straight ahead with empty, dead eyes and Bozo the Clown hair. The old engineer told us it was his saddest and most disturbing memory.

94

The boys were being treated by no fewer than six firefighters each. They were loaded on gurneys and quickly wheeled to the waiting ambulances parked out front. The side street leading to the yellow house was a throng of activity. As I escape the backyard, I saw the line of news vans with their satellite towers. It was just coming up on 5 p.m. and the local news channels were going to use this as their lead story. The well-dressed, pretty reporters, who were paid to stand in front of the camera and report tragedies, anxiously awaited any information they could regurgitate to their viewers. This, coupled with all of our apparatus, attracted a horde of onlookers. Traffic was bumper-to-bumper as the locals tried to get a look. Neighbors lined the street as kids rode bicycles back and forth in front of the house. I imagined the scene resembled a public hanging in the Old West.

I had done more than a dozen of these interviews during previous summers, although none of them had been for a double drowning. A horde of media set up behind the yellow hazard tape on the side of the house. They wanted to use the pool as a backdrop for the interview. I stood in front of the wall of cameras and gave a brief overview of what had happened. Several reporters asked relevant questions about how the boys accessed the pool, who was watching them, and how long they had been missing before anyone noticed. After talking for 5 minutes, the interview ended.

Most of the videographers had turned off their cameras, and the live newscasts had moved on to the next most depressing thing to report for that day. The crowd of reporters was breaking up when one of them shouted out the inevitable "How do you feel about this tragedy?" question. I looked at the guy who made the inquiry and told him I really didn't have an answer. Evidently, he didn't like my response because he immediately shot back, "What were your thoughts when you saw them scoop the second little boy out of the pool?" I stared at him for several seconds. To this day I don't remember if I said it just to myself, or if I shared my answer with the world. The only word I could think of to sum up the past 30 minutes was "motherfucker."

Our counselors had been on the scene for a few minutes and were dealing with the grandmother when mom got home. The ambulances had departed 5 minutes prior to her arrival, but several of the engines remained. She knew something was wrong. She bolted from her car and made a beeline toward the house. I wasn't up for the fourth act of this tragedy, so Dexter and I left after the counselors on our customer-care vans told us they would take the mother and grandmother to the hospital.

My first thought before we left the scene was to send everyone home for the rest of the shift. Sometimes this is the wrong course of action. Most firefighters have seen far worse than two intact, drowned children. Sending firefighters home oftentimes leaves them with nothing to do other than dwell on the fresh catastrophe. The members who have family at home aren't always better off. Thankfully, most people's families have never witnessed a tragedy. They can't always understand why their firefighter loved one comes home early and doesn't really want to talk very much. Most times, the best therapy is to stay at work and try to put the event into perspective with a group of people who have experience with these types of gruesome adventures — your co-workers.

On our way back to the station, I assumed both boys were dead. Engine 51 was back in quarters when we pulled in. We all sat around the dining-room tables, rehashing the event. Engine 44's crew stopped by to retrieve some of their EMS gear. They had treated the younger of the two boys and told us he was alive but on a ventilator. They said his older brother was pronounced dead at the hospital.

After a couple hours, our public information officer (PIO) called us. We filled each other in on what we knew about the incident. He told me the mother had just come from court. She spent the day divorcing her husband — the father of the 2-year-old who was still alive. Daddy was in prison, just like her first husband and the father of the dead 4-year-old. Our PIO said he was being inundated with media requests. He asked me if a few of us would talk to a local newspaper reporter. She had done him a variety of favors and wanted to run a feature story about the incident and its impact on firefighters. I told him why I thought it was a bad idea. I don't work with people who do very much emoting, and if you don't know us, you might even call us callous. The PIO assured me we would be a perfect group and requested that a couple captains and I do the interview. I told him he was crazy, but we would do it.

A family representative stopped by the station next shift to let us know the 2-year-old had died that morning. She brought the station a fire-truck-shaped cake to show their appreciation. We all exchanged condolences and suffered an awkward silence for a few minutes until the family rep said goodbye and left. The door had just closed behind her when one of the firefighters who had performed CPR on one of the boys walked over to the large, pink cake box. He picked it up and hurled it in a trashcan as he delared, "We don't eat dead kid cake."

The media flooded the airwaves with coverage about the incident and the pending funeral. Both boys would be buried in the same casket. Their fathers would get to attend the services wearing prison coveralls with their hands and feet shackled. They would be escorted by gun-toting corrections officers. I heard the two different sides of the family had begun fighting upon receiving the bad news. The funeral sounded like a swell time. Someone from the family called me and inquired if any of the firefighters who responded on the call would be attending the services. I told them no.

Our PIO showed up at 5 p.m. with a young female reporter. Paddy, the paramedic-captain from Engine 14 (the unit that worked on the 4-year-old), Dexter and I sat down to talk to the pair. The reporter was a young mother herself and wanted to know what effect these kinds of calls had on us. She told us if this event had happened to her, she didn't know how she could go on. The three of us told her we didn't enjoy going on these types of calls, but we had a job to do. If we had an emotional breakdown every time we saw something tragic, we would be very ineffective. The people who need our service call us because we show up quickly and take the appropriate action. If we were to show up only to become another set of crying spectators what good were we? I could tell the interview wasn't going the way she had hoped. Several times she countered us with, "Yeah, but what about…" She finally accused us of being cold when we implied the best firefighters were able to remain clinical when faced with human pain, suffering and gore. She kept going back to the emotional aspect of two little boys dying. Paddy finally summed up what the three of us were thinking in terms she could understand. He told her those two kids were innocents and didn't deserve their fate, but they were doomed. He would forever have the indelible snapshot of a 4-year-old boy caught in a pool net burned into his memory. She continued to look at him skeptically and implied that we should all be on the verge of an emotional breakdown. This pissed off Paddy. He was done with her pandering. "Where have you been?" he asked. Before she could answer, he told her he had been on four child fatalities in the last month. He then described an auto accident we had responded to a week before the double drowning. A mother and her 2-year-old girl had been killed while riding in the family car. Crews had to pull the dead mother off her only living child while the father watched a paramedic stick a plastic tube down his dead child's throat in a hopeless effort to make him breathe again. Paddy looked at her and said, "None of that was newsworthy, so no one asked us how we felt then. We've all been doing this a long time and have seen lots of nasty things. We don't need the media dictating how we should feel." We sat around and made uncomfortable chit-chat for a few more minutes, but for all practical purposes the interview was over. My trepidation to go along with this stupid interview was confirmed a couple days later. The story appeared in the Sunday paper with the headline: "Firefighters Trying to Cope With Tragedy."

Chapter 9
Fashion Suicide

During the course of my career, I have witnessed many examples of the ravages of heat. Heat gives life, takes life, and causes dead things to swell and explode. The crew of Engine 51 and I got to witness the devastating effects of storing dead human meat in an improperly climate-controlled space.

I was sitting on a chrome-framed sofa with green Naugahyde cushions while I studied for the upcoming captain's test. We had chairs that matched. This furniture was uncomfortable, and we affectionately called it "Big Green." I had been digesting my lunch of "Duke's Delight" (a derivative of goulash) for about an hour. It was early in the month of June, and the outside temperature was hovering around 110 degrees. Inside it was a comfy 70 degrees. The combination of lunch and the comfortable inside temperature was making me sleepy. I was sending a psychic thank you to whoever invented air conditioning. This man, whom I didn't even know, had done more for my quality of my life than any of the presidents of the United States. I slowly drifted off thinking about cold air and what it meant to me...

It's an unspoken rule on B Shift that during the day, the thermostat stays at 70. At night, we drop the thermostat another 10 degrees and keep the temperature in the high 50s to low 60s. Although we have no scientific data to back this up, it is well-accepted B Shift theory that you sleep more deeply and have better dreams at lower temperatures. The other two shifts don't subscribe to this hypothesis. I do not ever recall coming to work at a well-chilled station. A-shifters try to live in a perfect world. The temperature in their dreamland seems to be a constant 78 degrees. I'm sure it has something to do with maintaining enchanting hair and healthy gums. C Shift really doesn't have much choice in the matter, and that is our fault.

One of the side effects of maintaining B Shift's preferred temperatures during our scorching summers is the periodic freezing of the a/c unit. Evidently as the unit runs, it creates a certain amount of condensation. If the unit periodically shuts off, the condensation has a chance to dissipate. If the unit never cycles off, the condensation freezes. I have seen a/c condensers encased in blocks of ice the size of a small child. This is truly one of the oddities of living in one of the hottest and most arid regions in the world. Imagine standing on the roof of a fire station in Phoenix, Ariz., in the middle of the summer to find that an ice tumor has killed your a/c unit. It's enough to send a crew home sick. Lucky for me, this scene always plays itself out on C Shift after the B-shifters have all gone home.

Our proclivity for cool temperatures hadn't gone unnoticed. Our utility bills were a little high in the summer. We rationalized this by the savings incurred during the other months when we didn't use the heater. The city's Public Works and Facilities people tried to solve this problem in a variety of ways. Each one has met with failure. At one point, someone in Phoenix hired an environmental conservation engineer (whatever the hell that is). This man happened to be a very polite, intelligent, American-educated eastern Indian who went by the name of Hadji. One of Hadji's duties was to address the issue of regulating fire-station temperatures. Talk about shades of Big Brother.

Hadji met with the fire chief and assured him he could design a tamper-proof system that would maintain a constant temperature in the fire stations. The chief bought him lunch and wished him well. Hadji had chosen one of the newer stations, Station 76, to retrofit with his foolproof contraption. No station represented the classic shift stereotypes better than Station 76.

Hadji installed thermostats about 6 inches deep within the ductwork behind the air vents. All of these were hooked up to a computer that maintained the temperature at 78 degrees—perfect for the hair and gums. After Hadji's henchmen completed their work, he sat down and explained the new system. A Shift was fine with it, since that's the temperature they liked anyway. C Shift was moved to tears that someone cared enough to keep B Shift from freezing the a/c unit. B Shift told Hadji they thought this was inhumane, but they were more hungry than angry. They fed Hadji lunch and learned more about the system for a later attempt to bypass it. Hadji proudly explained how the system worked, tried to convince them of its merits, then finished his lunch and left.

About a week later, C Shift made an angry and hysterical call to report the a/c unit was blowing hot air. A serviceman went to the station to repair the system. Low and behold, when he took the cover off the unit, it was frozen. He immediately went back to his supervisor to report this impossibility. Word quickly got back to Hadji, who was incredulous upon hearing the news. He had engineered this system himself. He had a master's degree. How could this be? His world of order had been disrupted. Hadji told the fire chief that he had to meet with him about this matter. Once again, the fire chief bought Hadji lunch and listened to his story. The wise old fire chief told Hadji he would pick him up in two days (when B Shift was on duty) and take him by Station 76 to see if they could figure out what was going on.

Hadji walked into the station first and saw the appliance immediately. The fire chief almost giggled as Hadji shook his head and muttered a chant. Someone had built a shelf under the main air vent. A 500-watt floodlight sat on the shelf. A gigantic aluminum-foil funnel had been fabricated to direct the heat directly onto the "hidden" thermostat. A blowdryer attached to the shelf blew across the light into the vent. Hadji estimated it was moving 150-degree air across his strategically placed sensing device. Above the shelf, someone had taped a sign that read "Bedrock A/C Company." Hadji turned to the fire chief and said in his melodic eastern Indian accent, "Chief, I do not think I can control the temperatures in your fire stations." The chief patted Hadji on the back and the B-shifters at 76 fed him the finest lunch the city had to offer. That night, the B-shifters dreamed of rodeos where they rode 30-foot-tall Amazon princesses through oceans of blueberry pancake batter while playing tubas.

☠☠☠☠☠☠☠☠

My 10-minute siesta in the uncomfortable green chair at Station 51 came to an abrupt end with lights and a tone. A very pleasant female voice told us we were to respond to check an odor at a large apartment complex a few miles up the road. The four members of our crew boarded one of the most beautiful pieces of fire apparatus in the country and headed into the most intensely blue day I can recall.

Most unknown odors turn out to be sewer gas. Our standard course of action for this type of call is to verify that it is sewer gas and then pass the info to the water department. We were in no great hurry to get to the scene, so we went Code 2 (no lights or sirens). About 5 minutes later, we pulled into a large two-story apartment complex. It was less than a year old and was very well maintained. The maintenance man greeted us. The name embroidered on his shirt told the world he was Terry. He said he had smelled something heinous all day. We followed him through the pool area, where three or four mothers watched a dozen or so kids enjoy the water. We went up a flight of stairs to the second floor where Terry said the smell was the strongest. The odor was very powerful; it smelled like an animal that had been dead a long time. We asked if any of the tenants had pets. He proudly told us pets were not allowed.

The smell seemed to originate from one of four apartments. Terry said two male college students who were out of town for the weekend lived in one of the apartments. A man who was already at work for the day lived in another. A single mother and her two kids who happened to be frolicking at the pool lived in the third dwelling, and the fourth unit was vacant.

The five of us stood staring at each other when someone had a brilliant idea: Terry should use his master key to unlock the doors so we could search for the source of the awful smell. Terry took exception to this because he would need the tenants' written permission before entering their apartments. The captain told him swell; if that was the case, we would leave him to find the stinking thing on his own. Terry was only too happy to bend the rules this one time.

When he opened the door, odor exploded out of the college students' apartment. People describe the smell of death as sickly sweet. This more accurately describes a line of old ladies at a Sun City all-you-can-eat Sunday buffet, not rotting, metabolizing, putrid flesh. The smell is both indescribable and unforgettable. We gave serious consideration to going back to the rig and getting our SCBAs. We abandoned this idea when my captain called an ALS unit. He was certain someone was dead.

Anyone who smelled this way was well beyond the reaches of saving, so the request for paramedics was pretty much moot. Our captain was an old timer who had served about half of his career with the fire department before we ran EMS calls. He smoked a minimum of two packs a day. He believed the pound of bacon that he salted and consumed every shift counter-acted any ill effects caused by smoking. It was his philosophy that no one was dead unless a paramedic said so. This tactic also kept him from having to fill out what he called "those complicated goddamn EMS forms." You could braid the hair growing out of his ears.

Terry chose to guard the door while the four of us ventured into the young males' living space. We stood in the living room breathing through our mouths, filled with a sense of dread and impending doom. Twenty years earlier, I would have been very quiet and still, hiding under the bed from whatever apparition caused such a stench. The four of us went our separate ways, each searching a room of the two-bedroom apartment. The place was very sparse and surprisingly neat, considering two college guys lived there. I went into the small bathroom and found nothing out of the ordinary with the exception of mismatched towels. I willed myself toward one of the bedrooms while my three counterparts kept wailing, "Is anybody home?" I took the final tentative step through the stench and poked my head just inside the bedroom. An unmade queen-size bed with nightstands on both sides sat in the center of the room. A small TV was set up on cinder blocks in the far corner. A video game system was hooked up to the TV. There was a window overlooking the parking lot in the wall across from me, and light poured into the room through the slats of the mini-blinds. A severe case of heebie-jeebies kept me from actually entering room; I felt as if I were standing in the boogey man's shadow.

I went back to the living room and met up with the rest of the crew, and we quickly exited. When we got outside, my captain thanked Terry for standing guard and asked him to open the vacant apartment. It was laid out exactly like the one we had just searched. Neither the smell nor the sense of dread was as strong in this unoccupied unit. We all met outside again and moved on to the apartment whose owner was at work. This place was a pigsty. There was at least a week's worth of dirty dishes in the sink, on the table and in the bathroom. The trashcan overflowed onto the kitchen floor. The living room was littered with beer cans. A large plastic bong sat on the coffee table next to an ashtray full of marijuana stems and seeds. Dirty clothes were strewn everywhere. The bathroom was not a place one would want to be naked. It looked like a high-school biology experiment gone bad. Several different colors of mold seemed to grow in front of my very eyes during the few seconds I was in there. In the end, we found nothing but the signs of poor living. We all met outside on the balcony and decided we probably didn't need to search the single mother's apartment since she was down at the pool with her kids.

The ALS engine arrived to the scene. As they walked up, they commented that it smelled like something had died. No shit. We filled them in on the happenings prior to their arrival. We were about to go downstairs to search the apartments on the first floor when my captain decided we should take a last look in the apartments we had just searched. Terry opened the door to the college boys' pad, and I went back in for one final look around while the others went through the remaining two units. I went into the second bedroom the other firefighter had checked the first time around. Zip. I wandered back through the living room, into the kitchen, and then into the bathroom. While resisting the urge to unload the contents of my stomach, I marveled that a 4-inch wall separated this tidy bungalow from the filth next door. I was about to leave when I decided to take a look under the bed in the bedroom I had checked on during my first search.

I forced myself into the bedroom. I stopped at the foot of the bed and knelt down on the floor. I lifted the bedspread and peaked under the bed, fully expecting to come face to face with some horrible extravaganza. Nothing. I shook my head at my own foolishness, stood up and turned to leave. I was now facing the door. I screamed like a little girl. The wall contained a long closet that had a set of sliding mirrored doors. A large man was hanging against the wall. He was well over 6 feet tall, and his feet were about a foot off the floor. During my first pass through the room, I had only poked my head inside, never looking toward the corner of the room where he was hanging.

I tried to catch my breath, but my shrill screaming continued. I felt like I was losing my mind. The screaming subsided, dissolving into heavy sobs. I may have been going crazy, but I knew I wasn't crying. I looked over at a female firefighter from the ALS company as she ran out of the room.

I returned my attention to the rotting carcass. One of the college kids had wrapped a 2-inch nylon strap with a quick-release mechanism around his neck. It was looped around a hook he had screwed into the wall above the closet door. He wore a red satin teddy — it was a beautiful piece of expensive lingerie. Beneath it, he wore a very tight pair of purple velvet crotchless panties. His hands, which were handcuffed in front of him, held his green-and-black pecker in a death grip. His legs were covered with a pair of tattered fishnet stockings that reached midway up his large thighs. They were kept in place with a pair of red garters that had dice sewn into them. I still can't get over his shoes. He was wearing a pair of worn-out, white, paint-splattered high-top sneakers. Why go to all that trouble only to have the shoes ruin your outfit?

The guy had been dead for days. All of his flesh was mottled with streaks of purple, black, red and green. His shoulder-length blonde hair framed his heavily made-up face and long false eyelashes. The makeup gave a strange contrast to his discolored, puffy face. His wide-open eyes bulged an inch out of their sockets. His black, swollen tongue protruded from his mouth.

His large size was grossly exaggerated by bloating. After death, the body starts to consume itself. In the street, this is known as rotting. One of the by-products of this feeding frenzy is gas. If the skin stays intact, the body begins to swell. He probably had weighed 240 lbs. in life but appeared to weigh 400 lbs. at this point in his decomposition. The garters on his legs looked liked rubber bands squeezing water balloons. If the hanging dead guy had kept his unit chilled like our fire station down the street, his corpse would have made a much better presentation.

The rest of the posse quickly joined me. My captain, a longtime veteran of this neighborhood, looked the spectacle up and down and screamed, "I have got to get out of this shit storm of a first-due area! Everyone in it is a fucking career criminal or a self-destructive freak." He stared at me, shaking his head as he lit his 30th cigarette of the day. "You got that right," I replied. "Who in their right mind would wear those shoes with that outfit?"

104

The medics briskly determined there was nothing any of us could do for the patient. The police had been called to the scene and were doing their cursory investigation. They took pictures and interviewed Terry along with some of the other neighbors. It was evident what had happened. We are all just seconds away from death, and some people get a thrill by dancing right up to the point of no return.

Just before we left, a private ambulance arrived on scene. We were standing outside the door when the crew walked up with their gurney. One of them commented that it smelled like somebody had died. These guys were dressed in black polyester pants with crimson piping running down the leg. They wore long-sleeve crimson button-up shirts and gold five-point star badges. They looked like security guards at a Mexican whorehouse. Our salty old captain asked them if they had a body bag. They asked him why. He told them they would probably need it. One of the pimple-faced attendants told him they would make that determination when they saw the body. "We don't tell you how to do your job," he said, "so please don't tell us how to do ours." El Capitan's fun meter pegged. He told the misdirected youth, "It's all yours, son." The group of us left the scene in the capable care of the cops and the upstart ambulance crew. The rest of the story came to us a few hours later from two of the cops.

The two ambo brain surgeons went in and sized up the situation. They decide to move the bed out of the way, park their gurney in front of the hanging masturbator and lower him face down. The wonder kids were having trouble releasing the neck strap. Skin was sloughing off the dead guy's neck as they struggled to undo the collar. A quick-release collar is exactly that. It is designed for people who take themselves right to the point of unconsciousness before they ejaculate. You don't want to have to worry about clumsy knots when you're at the heights of self-induced passion. Verge of joy/verge of death requires quick-escape capability. The pimple-faced ambo jockey pushed the release mechanism, and the corpse fell directly onto the gurney like a crusty log of shit, just as they intended. What they didn't anticipate was for the corpse to blow up. And that's exactly what it did. The 5-foot drop was too much for the pressurized carcass. Rotting tissue, stagnant body fluids and guts exploded all over the two hapless know-it-alls. This was accompanied by a large release of stench more wretched than any other in the history of our blue planet. One of the ambo guys vomited directly into the soupy remains on the gurney. Both cops were forced out of the room when they looked down and saw one of those stupid tennis shoes on the floor. Rotten meat and a couple of lower leg bones sprouted from the top of the shoe.

On a professional level, I am very fortunate. I have certain work experiences that I will never forget. Various mental-health professionals are of the opinion that these profound experiences may contribute to a case of post-traumatic stress disorder. I beg to differ with this analysis, but 20 years after I participated in this event, I cannot look at a pair of worn-out, paint-speckled athletic shoes without flashing back to that fashion faux pas.

Chapter 10
The Toad & the Princess

It is quite common to watch humans disregard common safety practices in the name of convenience and laziness. Just the other day, we were driving down the road and saw a glowing testament to the lack of electrical current running through some people's gray matter. Three kids between ages 4 and 6 were playing jump over the seats while mom drove down the road at 45 mph. The children were so vibrant in their gymnastic activities that the car shook as it lumbered down the road. All of these kiddies fell into age and weight categories that require car seats, as mandated by the Society of American Pediatric Physicians, Divine Order of Orthopedic Surgeons, Fraternal Brotherhood of Emergency Room Physicians, Jews for Jesus, and the American Red Cross. This concept was lost on mom. None of the humans wore seatbelts, let alone car seats. Only one of the vehicle's occupants was securely belted in place: a 15.5-gallon keg of beer. It caught my attention when one of the kids sat on top of it. Mom was more worried about foamy beer than highway safety. A keg is a terrible thing to waste.

Any risk-management expert worth their salt will tell you that for the most part, our destiny lies in our own hands; it's a product of the decisions we make and the way we conduct our daily lives. If mom got into a serious enough accident, she and her kids would likely be injured or killed because they were not wearing seatbelts. The keg, on the other hand, might escape serious injury. Although we can improve our odds of survival, the mysterious hand of fate reserves the final say in our ultimate outcome. This was clearly demonstrated to me by two similar events that had drastically different consequences for the principals. They happened 10 years and 10 miles apart.

I was a firefighter working on Engine 7. It was the middle of summer and hotter than rat shit. My captain was a kindly old gentleman with an obsessive-compulsive disorder. He was very easy to get along with if you did your job. He had a proclivity to suck his teeth and demanded a tidy station. He was swimming against the tide with this facility. Station 7 was a dilapidated two-bay firehouse that had been built in the early 1950s. If you ran the water in the kitchen sink too long, the urinals in the bathroom would overflow. The pipes shook so hard when you flushed the toilet that plaster would fall from the ceiling. They don't build them like they used to.

Engine 7's engineer was a man named Ralph. He was very secretive, and some of his co-workers began to doubt whether Ralph was really his name. He was better known as Rocket Ralphie the Race Car Boy because of the way he drove Engine 7. He knew all the shortcuts in our 5-square-mile first-due area. Ralphie studied hydrant maps for hours at a time and knew the location of every fire hydrant contained in the same region. Our station was about two blocks from a main intersection, and based on the color of the light at that intersection, he could tell you how fast he needed to go to get green lights at the intersections in every direction. "If the light at 15th Street is yellow, I must go 58 mph to catch a green light at 7th Street. If we are heading west, I have to go 62 mph to get a green light at 19th Street." Ralph went as far to keep a minimum amount of fuel and tank water in the rig. The extra weight slowed him down. The rig's booster tank held 500 gallons of water, which weighs about 2 tons. At the start of the shift, Ralph would dump some of the water out of the tank. He was going to get you on scene first or crash trying, which he did with some regularity. He once told me it was better to go through red lights at a higher rate of speed because it reduced your exposure time. Ralph had different thought patterns than the rest of us. For example, he felt our 20-year pension should be changed. Instead of working one day out of three, he reasoned we should work 24 hours a day, seven days a week for a period of seven years. He surmised we would make triple the salary and retire with three times the pension. He argued that if prison inmates could do it, then so could we. Ralph sincerely thought this way because he was out of his mind.

Station 7's first-due area was full of winos and prostitutes. Every B Shift Friday, Paramedic Joe of Rescue 7 gave the winos haircuts in the apparatus bay. He soon began coloring hair as an added service. It was not uncommon to see eight or nine winos walking the streets with lime green, stark red or neon blue hair. They all looked as if they felt a little better about themselves. They all loved Paramedic Joe.

One afternoon around rush hour, we got a call for a car vs. pedestrian at the intersection just south of the station. Traffic was heavy. All the down-town workers were in a rush to leave the core of the city, abandoning it to the night shift of humans who slept in its parks and alleys.

The accident was just east of the intersection, and it had stopped traffic in all directions. When the rig came to a stop, I jumped off to grab the EMS gear. As I stepped onto the road's soft, molten asphalt, I noticed the normally lethargic crowd of street people was quite excited. Some of them were even screaming and beckoning for us to hurry. Large thunderclouds were moving in from the south. They added stifling humidity to the 115-

degree heat. I felt like a prisoner trapped inside a gigantic, angry vagina. The crew was exchanging funny glances with one another as we made our way through the throng of hysterical onlookers.

These people lived harsh lives and had seen many of their friends meet bad ends to their fractured lives. Most of them didn't complain or bitch; they learned it didn't do much good when they did. Each consumed their own special tonic that evened out life by frying the proper receptors in their brains. This special group of social rejects called the fields, parks, alleys and underpasses in this stretch of town home sweet home. Many of them refuse to live in the myriad shelters and missions the valley has to offer. They find the rules too confining. They have apathetically accepted their miserable lot in life and nothing ever seems to upset them. They are hard people.

The current situation had them acting like chimpanzees being stalked by lions. As we broke through the mass, the first thing I noticed was a well-dressed woman standing beside a new SUV with a damaged front end. She was screaming at the top of her lungs, "Oh no! Oh my God! Someone please help!" Despite her expensive makeup, she was very pale and beginning to sweat. Rings of perspiration were forming in the armpits of her ivory-colored designer silk blouse. Tears streamed down her cheeks.

Wedged under the front of the large pearl-white vehicle was a mangled wheelchair. The chair belonged to Sonny—we knew him by his street name, Bunghole George. Bunghole was a Native American and one of God's children. I had first met Bunghole several years earlier when I worked at a downtown fire station. We had been dispatched on a "man down" call. When we arrived to the scene, we found a one-legged man lying facedown in the middle of the street. Gus, the senior firefighter on the rig, told me to stay away from him. From the safety of the sidewalk, Gus yelled, "Get up, Sonny. We're not going to let you do the nasty to us. If you don't move your ass out of the road, I'll throw flares on you." Sonny was grumbling some type of incoherent drivel as he rolled over. The man was quite a sight. Not only was he was missing one of his legs, but one of his eye sockets was little more than a black hole. In his filthy hands, he clutched his colostomy bag. His tactic was to lie face down and wait for a Good Samaritan to come to his assistance. He would then roll over and blast them with the contents of his bag. Some said Sonny lost his leg to diabetes, others claimed he had fallen asleep on the railroad tracks. Sonny wouldn't say. He was pissed that his fun for the day had been ruined. Sonny's buddies magically appeared from an alleyway, dumped him back into his broken-down wheelchair and pushed him down the road.

Bunghole had suffered many of the traumas the world has to offer. Virtually all of his injuries were the result of bad decisions. He has been shot and stabbed (which is how he lost his eye) on several occasions. One night, while he was enjoying a drink around a bonfire with some of his friends, he passed out and fell out of his chair, face down into the fire. Several minutes passed before Bunghole's peers noticed him and pulled him out of the cozy campfire. When A Shift got to the scene, his hair was still smoldering, and one of his ears had been completely burned off. The crew figured that was the last time they would see him. A couple weeks later, he was being pushed down the street looking like a mummy. He was immortal. I was convinced that within a few short years, Bunghole would be nothing more than a blind, disfigured head rolling down the street, spitting at people.

The impact of the collision with the large SUV knocked Bunghole out of his chair and about 25 feet down the street into the large curb. He was screaming, "That white bitch ran us down." He was bloodied but appeared relatively intact. Bunghole's personal assistant, a tall transvestite who went by the name "Ho-Ho," did not fare as well. Ho-Ho stood a few inches short of 7 feet tall. He weighed in around 160 lbs., and his rail-thin body was capped with a heavily pockmarked face and thick, curly black hair. His bucked teeth accentuated a large, grotesquely bulbous nose. Some nights, he could be seen wearing heavy makeup, 6-inch heels, a sequined cocktail mini dress and a long, platinum-blonde wig. He was less attractive as a woman than he was as a man. It was quite a sight to watch Ho-Ho push Bunghole around when he was all gussied up. The toad and the princess.

One's personal safety is generally at greatest risk when you're not thinking clearly. The leading cause of repressed thinking is the intake of intoxicating substances. Bunghole and Ho-Ho were always high on something. That's why they met up with the high-velocity front end of a large American gas-guzzler. Their favorite potion was Thunderbird, a very powerful and relatively inexpensive wine.

Ho-Ho would ask Bunghole, "What's the word?"

"Thunderbird."

"What's the price?"

"50 twice."

The spectacle that had the crowd so amped was Ho-Ho. The SUV had hit the tall transvestite just above the knees. The impact snapped his legs clean in half, leaving them with open femur fractures. Most people with open leg fractures don't move around very much. This wasn't the case with Ho-Ho. The large bone ends were sticking out of his chewed-up lower thighs. They glistened like an A-shifter's tooth enamel. Ho-Ho was balancing on his hands and "walking" toward Bunghole. He would pick up one of his broken legs, move it forward and plunk it down. Every time he did this, the fractured bone would impale itself several inches into the hot asphalt. He would then take another step and plunge his other broken bone into the hot street. He looked like a man walking in deep snow, only he was dragging about 4 feet of knees, lower legs and feet—adorned in cork platform sandals—behind him. Meanwhile, Bunghole had started spraying the crowd with the contents of his colostomy bag. It was both touching and nauseating.

We subdued Ho-Ho, splinted his ruined legs, tied him to a long backboard, and performed all the other required treatment protocols for someone with several cups of hot asphalt on the shattered ends of their body's longest bones. We all figured he was as good as dead. Even if he survived the ordeal, he was certain to get infections in both legs. Chances were that if he lived, he would end up loosing both limbs.

Ho-Ho didn't make it out of the hospital. A combination of cheap alcohol, brisk traffic and a disregard for his personal safety took him out. He was probably too beautiful for this world anyway.

Ten years later and about 10 miles to the east, I went on another car vs. pedestrian accident that got me thinking about Ho-Ho's trauma-shortened life. When we got to the scene, we found all the elements of your normal automotive tragedy. An older station wagon was parked cock-eyed in the middle of the street. A pair of women's orthopedic shoes sat in the center of a mid-block crosswalk about 40 feet behind the stopped car. Empty shoes in the middle of a busy street are always a bad sign.

I hate mid-block crosswalks. It doesn't make sense to put a crosswalk in a place where traffic is usually moving at 50 mph. Phoenix is laid out in 1-square-mile increments. Traffic lights are usually placed every half-mile, with additional crosswalks every quarter mile or so. There are no warning devices or traffic lights to remind drivers to slow down for pedestrians at these quarter-mile crossing points. The fatality rate for pedestrians hit in these crosswalks is higher than the death rate experienced by extreme skydivers. People who use these crosswalks are throwing their personal safety to the wind.

We parked our rig behind the station wagon that had just nailed the pedestrian. As I walked around to survey the front of the vehicle and search for the patient, I saw the car's windshield was shattered and caved in. A large wad of gray hair remained wedged in the shards of broken glass. This sent us looking for a dead old person who was missing a large amount of their scalp.

The driver of the car stood on the curb arguing with one of the bystanders. The bystander claimed the driver was at fault; the driver was countering in some kind of Eastern European dialect no one understood. The only part of her explanation that made any sense was her overuse of the phrase "cocksucker." An old woman holding onto a bent-up two-wheeled grocery cart stood next to the angry driver. I approached the bystander who was arguing with the driver and asked if anyone was hurt. He pointed at the old woman and said she was the one who had been hit. Before I could argue, I looked down and noticed granny was shoeless, but she didn't have a scratch on her. I asked her what happened. She said she was crossing the street, minding her own business, when "that crazy bitch hit me. It's bad enough these foreigners come to our country and suck the life out of it. Now they won't be happy until they kill every last one of us. This country has been going to hell ever since the Kennedys. As far as I'm concerned, Oswald was a patriot."

Before granny got more carried away about our country's foreign policy and dead Presidents, I cut her off.

"Let me get this straight: You were crossing the street, and that station wagon hit you."

"Yes it did. That Commie bitch tried to kill me but Jesus pushed me out of the way."

While we were having this conversation, the rest of our crew surveyed the car. Luke was on his hands and knees, looking under the thing to make sure it didn't contain any wadded-up human surprises. I was scanning Granny's hair to look for the missing clump that was wedged in the windshield when Jake walked over wearing a gray wig. Granny giggled and asked him for her polyester hair.

The bystander told me the car was going between 35 and 40 mph when it hit Granny. The old woman countered, "Bullshit. She was going at least 70 when she tried to run me down."

The bystander ignored the old lady and said, "The old woman went up into the windshield head first and then tumbled over the top of the vehicle and landed in the grass on the side of the road. It looks like her wig took the brunt of the blow and saved her life."

Granny refused to go to the hospital to get checked. We were all standing around waiting for the cops to show up. I hadn't thought about Ho-Ho for the past decade. I don't know why the image of him plunging his shattered femurs into hot asphalt flooded my mind. I mumbled to myself, "Why didn't Jesus push Ho-Ho out of the way?"

Granny gave me a stern, penetrating look before saying, "Son, if you know what's good for you, you'll never second-guess Jesus."

Ho-Ho and Granny went to the same dance and only one of them got to go home. The old lady took off down the road, pulling her dented cart full of bruised melons, with a crooked wig and a renewed sense of disdain for her fellow man. Both Granny and Ho-Ho broke the rules of personal traffic safety with wildly different results. But what about people who follow the rules and end up being blown to bits by fate? Life can be confusing and unfairly brutal.

☠☠☠☠☠☠☠☠

Our job brings us into direct contact with life's losers. In most instances, when life reaches across the table and bitch slaps your head from your neck, it's because you broke one of her rules. People who thumb their nose at nature (life's mother) are cool right up until the point that velocity and gravity grind them into pink, frothy pulp. Insurance companies finance large buildings full of risk managers to determine how much it will cost to insure individuals. Much of this assessment is based on how much, and how often, the customer tempts fate. The insurance industry uses mathematical formulas and statistics to determine the insuree's likelihood of death. Based on their calculations, Bunghole George was uninsurable. On the other hand, a family having a Labor Day backyard barbecue would be considered a very low risk. If only life were that simple.

I was working as the captain on Engine 53. We were in the process of telling a homeowner that he wasn't allowed to burn the plastic jacket off of copper wire. This is a common practice in economically depressed neighborhoods but hasn't caught on in the more republican areas of town. Mr. Homeowner was giving us grief over hassling him. We pointed out that this salvage practice produced a large amount of offensive smoke that bothered the other folks on his block. We were making lovely progress toward a positive customer-service outcome. As he was putting out his fire with his trusty garden hose, the gentleman looked to the horizon and said, "Ain't this something. Here the fire department is out hassling me, a working man just trying to scrape together a few more dollars to feed my family, and over yonder is a real fire."

We looked to the north and saw a very large column of black smoke about a mile away. We quickly left the backyard and made our way to the rig and headed toward the conflagration. I was frantically putting on my turnout gear along with the two other firefighters, Paul and Fred. Max, the

113

engineer, was pushing the rig to warp speeds. I had cleared the alarm room to report the smoke's general area and to tell them we were heading that way. The alarm room operator told me to "standby," which is just a nice way of saying, "Shut up, I'm kinda busy with more important things right now." Being a team player, I shut up. A second or two later, an Alert 3 (a reported airplane crash) was dispatched at Sky Harbor Airport, which is just a few blocks east of Station 53. These alerts all end the same way: The dispatched units rush to the airport only to watch an airplane safely land and taxi to its gate. I looked to the south in the direction of the airport and saw nothing out of the ordinary. After the dispatcher finished spitting out the airport call, I got back on the radio to tell her about the smoke. She came back and told me to go available; the source of the smoke was probably a house fire that already had units on the scene. We were heading down 44th Street and could see fire above the rooflines of the houses several blocks away, but we still hadn't heard the first units arrive. I came back to her with my own "stand by." As we made a left onto Flower Street, I could see Engine 10 laying a supply line to the burning house. They gave their on-scene report, "Engine 10 on the scene of a working house fire. We have heavy fire and smoke to the rear of the house. We're laying a line in and attacking with an inch and a half. Engine 10 will be command."

As soon as Engine 10's captain finished his initial report, I reported that we were staged to the east. He ordered us to pump the hydrant, pull a second line off Engine 10 and take the inside of the house for search, rescue and fire control.

Max parked at the hydrant while the rest of us got off the rig, threw on our SCBAs and pulled a second inch and a half from Engine 10. While my firefighters stretched the line, I could see Engine 10's crew working their line. They had knocked down the fire in the carport and on the side of the house. The front end of the car parked in the driveway was still smoking. A workshop had been added to the back of the carport. It was as wide as the carport and as deep as the backyard. Heavy fire and smoke were pushing out its roof. Engine 10 was knocking down the fire around the gate that led from the carport into the backyard when a loud explosion scared the shit out of all of us. One of the tires on the smoldering car had popped. In the second it took to me figure out what happened, I could see Sue, the firefighter working Engine 10's line, smile at me through her SCBA face piece. As I went back to the front of the house, I saw a small group of people in the neighbor's front yard. Some of them were hysterical while others were catatonic. A few of them were bleeding.

I hooked up with my crew as they prepared to take the line inside the house. Paul had the charged line in his hands and was patiently waiting for Fred to kick in the locked front door. The opened doorway immediately filled with smoke. The three of us got under it and crawled inside. The smoke banked off the ceiling. From our hands and knees, we could see only the bottom 4 feet of the house. We did a quick search of the living room and could see the fire had burned through the Arcadia door to the rear of the house, and it was free burning in the kitchen and dining room. I communicated with grunts and hand gestures to Paul to keep the fire from extending into the rest of the house while Fred and I searched the two small bedrooms. I went to the little girl's room to the right; Fred took the parents room to the left. As I crawled from the hallway into the bedroom, I admired the smoke as it swirled, rolled and folded over and into itself while it banked farther down the walls.

Smoke shares a lot of characteristics with cotton candy. The biggest difference being that most smoke ranges from grey to black while most cotton candy ranges from pink to blue. The shared characteristics change considerably when smoke becomes heated. Hot smoke looks like thunder clouds captured on time-lapsed video. Much like cloud-watching, if you watch smoke long enough, you can make out shapes such as dragons, Snoopy, elephant heads and other images we remember from the lazier days of our childhood.

Smoke carries a variety of downsides for all lung-breathers dependent on an uninterrupted supply of fresh air. First and foremost, smoke is unburned fuel—comprising mostly unburned particles of carbon—that didn't get hot enough the first time it burned to actually ignite. Smoke is a means of transport for a lot of different fire gases, including carbon monoxide, hydrogen, ammonia, toluene and cyanide. Many of these gases are toxic and flammable, and just like their carbon cousins floating in the blanket of smoke, they will burn when the fire makes things hot enough. It can be hypnotizing to watch black smoke glow red from its center, just like lightning illuminates the inside of a storm cloud. This is a sign that all the right things have come together in the proper amounts—there's enough heat, oxygen and unburned fuel for ignition. When this happens, the smoke becomes fire in the blink of an eye.

A lot of crazy stuff can happen after smoke becomes fire. If the transformation takes place in an area that doesn't have a lot of combustibles, the smoke can quickly burn off, and the process begins again. On the other hand, if the area is full of furniture, mattresses, carpeting, doll babies and the other paraphernalia one would expect to find in a little girl's bedroom, the contents of that area burst into flame.

Another smoke-related hazard is that it is impossible to see through with the naked eye. From their earliest training, firefighters are taught not to leave the protection of a hoseline when they operate inside a burning building. There are a few exceptions to this rule. The group of us entered the burning house with an attack line, which was currently babysitting the main body of the interior fire. This tactic was buying us the few precious seconds we needed to search the bedrooms of the small house. I was in the process of making a right-hand wall search of the bedroom. Keeping the same shoulder against a wall ensures you don't get turned around and causes you to exit through same opening you entered. The smoke was now only a couple of feet off the floor, and it was getting hotter. In a few seconds, the room would be fully charged from floor to ceiling with pressurized, super-heated black smoke. I was crawling on my hands and knees, sweeping under the bed and in the closet, following the wall and the layout of the room. I could still see a foot off the floor and quickly finished my work in the small bedroom and met up with Fred in the smoky hallway. It took us less than a minute to search both rooms. We checked the bathroom on our way back to hooking up with Paul. I radioed Engine 10 and said we had an "all clear" in the house. He came back and told me they had heavy fire in the backyard, and it had extended into the attic of the house.

Paul opened up the line and hit the fire in kitchen. The little bit of visibility we had quickly turned into ghosts of white steam, blurs of brown smoke and intermittent flickers of orange fire. I heard Engine 10 order Ladder 11 to cut a hole in the roof of the house. When Paul had most of the fire knocked down, Fred used his pick-headed axe to open up the ceiling so we could put water into the burning attic. The ceilings in the old house were made out of lathe and plaster (strips of 1-inch lumber nailed to the ceiling joists, covered in wire mesh then frosted with plaster). These ceilings are very difficult to open. Paul directed his hose stream into the burning 2-foot hole Fred had pulled when the fire kicked back up in the kitchen. I told Paul to forget about the attic fire; we needed to knock the fire down in the kitchen then move out back to help with the sheets of fire that seemed to be appearing all over the backyard. We killed the stubborn fire in the kitchen, then we went through the melted Arcadia door to battle the outside conflagration with Engine 10.

The workshop was to our left. A large bottle of acetylene gas was venting. The escaping gas made a high-pitched whistling sound while fire jetted 20 feet out of the bottle's top. Fire rolled across the entire back of the house up into the eaves of the roof. The backyard pool was full of muck and debris; the water level was about a foot too low. Fire was everywhere.

It didn't make any sense. Every time Paul knocked the fire down in one place and moved his line, the fire would kick back up. It was as if someone had dumped gasoline on everything.

While we had been inside the house, Engine 10 had balanced the fire to a 1st alarm. Engine 11 quickly joined us in the backyard while Engine 12 and Ladder 12 remained inside the house to deal with the attic. We were advancing on the fire in back when Paul and I fell into a large hole filled with water. The hole was almost 3 feet deep. As we climbed out, we could see a picnic table at the back corner of the house. Sitting on the bench, with his back to the house, was a severely burned corpse. Both arms were outstretched and burned off to the elbows. The legs were burned off to the knees. He looked like a quadruple amputee. We made our way to the corner of the house to extinguish the remaining fire burning against the back wall. Tony, Engine 11's captain, was mesmerized by a small piece of torn orange aluminum. Ladder 12's crew remained on the roof and reported to command that they had found part of an airplane wing on the roof.

There were about 10 of us in the backyard, and most of the fire was knocked down. I turned my attention to the pool. A thick blanket with a lump underneath it lay half on the deck and half in the dirty pool water. I was picking up the blanket when one of the firefighters told me I shouldn't. He was too late. The blanket was covering a little girl who was missing her head and most of her torso. Deciding the pool probably contained more grizzly mysteries I didn't want to discover, I went over to Fred and Paul.

It was quite amazing. There were at least 15 firefighters in the backyard. No one spoke. We seemed to realize simultaneously what had happened.

It was Labor Day weekend. The family who lived in this house did not want to hassle with the roads and throngs of people who leave town for the three-day weekend. Instead of dealing with all the stress and the drunken drivers, they opted to have a backyard barbecue. Mom and Dad, two little girls and a couple of neighbors were just kicking back and enjoying their day. The lawns were manicured, and everyone was at peace. At the same time, a small airplane took off from the airport 5 miles away. Something went wrong, and the plane fell into a steep dive. We were later told the plane impacted the backyard at an estimated speed of 250–275 mph. Two of the people in the backyard saw what was about to happen and ran for their lives. They were the bleeding patients we saw in the neighbor's front yard. The plane had impacted the backyard barbecue right next to the picnic table. When the small plane impaled itself into the ground, the 100 gallons of fuel in its torn tanks covered the entire backyard and burst into flames. The two little girls had been floating in the pool on a raft. A piece of high-velocity wreckage tore one of them in half before she knew what was happening. A seasoned police forensic photographer broke down after having to photograph the ruptured remains in the bottom of the pumped out swimming pool. Life and death can be real motherfuckers.

Fate is life's wild card. Bunghole George led a life that would kill a small village of God-fearing Christian folk, yet nothing could stop him. An all-American family was wiped out in their own backyard by a small airplane. Granny's life had been spared so she could continue to radiate negative vibes toward her fellow man and wear cheap helmets of polyester hair. Don't try to make sense out of this. It will only make you crazy.

Chapter 11
Big Rex & the Cookie Toss

Sometimes it seems odd that people consider firefighting one of the best jobs on the planet. We see more than our share of blood, guts, destruction and human misery, yet more than 90 percent of us have careers that span beyond 20 years.

I was 20 years old when I started my career. Before then, I had never seen a dead person. I hadn't even been to a funeral when I began the training academy. There were two phobia tests we had to pass before we could graduate and become firefighters. The first was a claustrophobia test. While wearing all of our firefighting gear, a canvas sack was placed over our heads. We then had to follow a hoseline into the smokehouse, which had several fires burning in it, and follow the line in total darkness as we navigated our way out. To complicate the test, mattresses, sofas and other large pieces of furniture obstructed our path. This has been the most realistic job-related test I have ever gone through. It taught me never to leave the safety of my hoseline. The second test was the acrophobia drill. This test is rather dangerous and only determines whether the candidate is afraid of heights. We climbed a 100-foot aerial ladder to the tip. When we reached the top rung, we hooked the D-ring on our ladder belt to the top rung of the ladder and leaned back, which is a really stupid thing to do if you think about it. No one in their right mind would climb seven stories above a concrete parking lot and lean back, trusting an old nylon ladder belt with their life. The first guy up the ladder leaned back so far that his helmet came off his head. The fall proved fatal for his manly hat. The test was pretty much pointless and served no real purpose other than to fatten the coffers of fire-helmet manufactures and prove you could climb a 100-foot ladder. Neither of these tests had me worried, but I wasn't looking forward to our trip to the morgue.

Our group visited the county morgue on a Monday morning after a very busy weekend of death and destruction. The chief coroner was an older gentleman with white hair and a goatee. He always had a cigarette hanging out of his mouth. His assistant was an old Phoenix Police Department detective. The pair explained the procedure of processing a dead body. Most of their talk was cop related. Fingerprints, entrance and exit wounds, time of death, as Joe Friday would say, "Just the facts." They explained that in the fire department's zeal to save people, we oftentimes destroyed crucial evidence, and we should think twice about trying to revive obviously dead people. The captain-paramedic chaperoning our group took exception to justice taking precedent over life-saving efforts. The discussion ended when the cop started wheeling dead people out of the refrigerator.

Although I had spent the first 20 years of my life without seeing the finality of death, within my first 10 minutes at the morgue, I was surrounded by 15 corpses. The first was a 20-year-old blonde female who had been killed in an auto accident. Next came a 16-year-old who could have passed as her brother. He had been shot in the chest at close range with a shotgun. Before I knew what was happening, I was surrounded by old women who had died in their sleep, a middle-aged Chinese man who had died during heart surgery, junkies who had died from heroin overdoses, and several losers from the Saturday Night Knife and Gun Club. My classmates and I were trying to act as if this was something we did every Monday morning. The coroner and the cop couldn't have cared less. Things got crazy when they started cutting up a biker who had been shot multiple times with a small-caliber handgun. The cop used a scalpel to make an ear-to-ear incision around the dead guy's head, and then he pulled his scalp over his face—literally pulling the hair over his eyes. As if this wasn't funky enough, he picked up a small bone saw and removed the top of the dead guy's skull. While he was carving the skull, Mr. Coroner made a deep incision from the guy's pubis bone all the way to his throat. This gouge was finished with a Y cut that spanned from one shoulder to the other. After the good doctor finished his incision, he pulled the dead guy open, used what looked like a set of tin snips to cut the sternum from the ribs, and pointed out the contents of his shot-up cavities. As he was showing us the guts, he took his scalpel and used it to puncture the guy's bladder, sending a fountain of urine everywhere. It was the only time I heard him laugh or show any joy. Meanwhile, the cop had removed the biker's brain and was filleting it with a bread knife. For the next hour, we got to pass around the dead biker's organs.

That morning's activities knocked the edge off of the whole blood-and-guts part of our new jobs. Our morning at the morgue prepared us for what we would see and deal with for the next 30 years, but by no means did it immunize us against the trauma we would witness during the course of our careers. I have witnessed some pretty medieval and bizarre things while on the job. That is one of the exciting and mysterious aspects of a firefighting career—one never knows what to expect.

One of our job's advantages is the limited amount of time we actually spend delivering service. For the most part, our encounters are pretty brief. Other than really big fires, the vast majority of calls last less than an hour. Traditionally, we show up after the first act of whatever tragedy had just occurred, stabilize the situation, then we hand the customer to whatever group can best meet their needs. For medical calls, this is the ambo crew that will transport the customer to the hospital. If the customer had a fire, we do our thing then hook the customer up with their insurance agent, the Red Cross or one of our social-worker types. If the customer is hearing voices or has telephones ringing in their stomach, our social workers transport them to a quiet, dark, soundproof place. We do not enjoy a long-term relationship with the customer, nor do we have to deal with the baggage involved with these types of associations. This makes our job much better. It's like only going out on first dates.

For most events, we usually respond before knowing what led to our dispatch. This has both good and bad points. The good point is that it makes it much more difficult to be judgmental or assign blame for whatever tragedy just occurred. Oftentimes, it takes the police days to reconstruct exactly what happened and to learn who was at fault. The downside of not knowing is just that—not knowing. Sometimes the circumstances surrounding the event are just too juicy or strange not to get some answers. We are human, after all, and are just as curious as anyone else.

One day we were dispatched to an "unknown medical" call. This is how our alarm room dispatches medical calls when they don't know what the problem is. We arrived on the scene to find a woman sobbing uncontrollably in her front yard. A total of six of us responded: Four firefighters on an engine company and two paramedics on a rescue unit had been sent to help this woman with her unknown medical condition. We did what any group of men would do for a crying woman—patted her on the shoulder and told her it would be all right. She knew we were full of shit, but she was either too upset or kind to say anything about it. After several minutes of being a caring nurturer, I began to wonder why this crying woman was holding a plate of cookies. I started to wonder if she would mind if I had one of them. She finally calmed down enough to tell us she was OK, but we needed to check the woman inside the house.

About once a week, the crying lady (I don't know what else to call her) would visit her elderly next-door neighbor. She would bring a tray of cookies, granny would make a pot of tea, and they would sit and visit for a while. On this particular day, the crying lady did not get a response when she knocked on the front door. The door was unlocked, so she let herself in. I can see her now, calling out the old woman's name in an effort to make sure everything was all right. She got no response as she wandered deeper into her neighbor's home. When she reached the bedroom, she found granny. Evidently, the old gal had been sitting on her bed getting undressed when she suffered some type of life-ending medical malady. When she died, the old lady was wearing a brassiere, a smart blue wool skirt, white hose and orthopedic shoes. She had collapsed straight back onto her pillow with her feet on the floor. I assumed she had died immediately. But it wasn't her death that had the crying lady out of sorts.

The old woman did not have one of those "I went peacefully in my sleep" looks on her face. In fact, she did not have a face at all, only two giant blue eyeballs and a puff of frizzy gray hair—as if she had taken her hair down and fluffed it vigorously right before she keeled over. All that remained of her face were skeletal bones, with very little skin or flesh on them. The rest of her body seemed pretty much intact; she wasn't decomposed or bloated. It was freaky. As the six of us stood there, trying to figure out where her face had gone, we heard a small bark from under the bed. Bad dog. Granny had died, and over the course of a couple of missed meals, little Fluffy became hungry. Fluffy couldn't figure out why her owner wasn't getting out of bed to feed her. Fluffy tried to wake her owner by licking her face. This licking progressed to eating. Once a small lap dog tastes human flesh, it cannot be trusted around people ever again.

None of us was upset about what happened to Granny. The old gal no longer had any use for the sack of skin and bones she once resided in. It's not as if the obnoxious little dog ate her alive. We did all we could do for the crying lady neighbor, which amounted to telling her that her friend had passed away while sitting on her bed. She felt no pain and had been violated by Fluffy long after she met her maker. This helped a little bit, but I'm sure she still has those nasty images pop into her mind every now and then. She also probably remembers how kind the "nice firemen" were to her. We became a temporary outlet for her cookie distribution after granny checked out. People rarely make cookies for people they dislike or can merely tolerate.

☠☠☠☠☠☠☠☠

In general, firefighters are very tough people who don't let much bother them. This doesn't mean I haven't worked with more than a few comrades who occasionally get queasy. One of the most sensitive of these people was a big paramedic named Rex. If a patient made retching sounds like they were going to vomit, Rex had to make a choice—leave the area or stay and blow chunks.

Rex was tall and had a big belly that often protruded from under his uniform shirt. He had narrow blue eyes and dressed like a cowboy on his days off. Rex seldom ate with the rest of us because he was always on a diet. One day around lunchtime, Rex sat eating more than 2 lbs. of mashed potatoes covered with an equivalent amount of melted butter. He was sitting at a table by himself watching Smurf cartoons. A waterfall of melted butter poured off his plate, onto the table and down to the floor. (I am not making this up.) I was sitting one table over with Sean, the other firefighter on Engine 23. We were wondering at what point mashed potatoes with lots of butter became diet food when the lights came on and the accompanying voices told us we had a code at the city building. We slid the poles, mounted the rig and fought the downtown lunchtime traffic as we drove to the city building. This is where people go when they want to get permits to construct buildings. They get to meet building inspectors who are more than happy to give them good customer service and help them build their dream homes. (I am making this up.)

A woman met us at the curb and told us an employee had been eating lunch when he just fell out. We grabbed our EMS gear and went inside to save the day. When we got to the second-floor cafeteria, we found a large unconscious man. He was more than 6 feet tall and weighed at least 400 lbs. He had a sparse beard and wore a large wristwatch with a hideous silver-and-turquoise Indian watchband. His pasty-white, clammy skin was drenched with perspiration. He was quite dead. The paramedics quickly sprang to action. The rest of the crew helped. Rex, with mashed potatoes on his shirt, began intubating the patient by sticking what looks like a skinny steel shoehorn with a flashlight on its end down the guy's throat. Once he got it in just the right place, he stuck a plastic tube down the man's trachea, allowing us to pump fresh oxygen into the lungs. While this was going on, one of us performed chest compressions. We put our hands on the dead guy's sternum and drove it into his heart. (Most of the free world has seen this quite a few times on TV.) The other medic, a man named Ernie who had a little Dutch boy haircut and could speak Russian, had started an IV and was administering cardiac drugs. A well-oiled, just in the nick of time life-saving machine. No wonder the public loves us.

Once the ambulance arrived, we loaded the large dead guy on the gurney and headed down the stairs because he was too long to fit in the elevator. The dead guy's feet hung a good foot over the end of the gurney. We finally made it out of the building and loaded him into the back of the ambo, which was doubled-parked on a very busy thoroughfare. Cars were squeezing by at reduced speeds because their drivers had a burning desire to see what was happening. The gurney that barely held the big dead guy was securely locked into the back of the ambo. One of the ambo guys was doing chest compressions while Ernie, the paramedic with the little Dutch boy haircut, was squeezing oxygen into the dead guy's lungs. That's when the miracle occurred. Sean and I were standing just off the curb to protect ourselves from the distracted drivers trying to get a better look at any decapitated people (much of the motoring public associates ambulances with headless corpses for some reason). Rex was stepping up into the back of the ambo, showing about 4 inches of his ass to the world, when the big dead guy's hand grabbed the oxygen bag and tube Rex had expertly inserted down his throat. He pulled the entire apparatus out of his body as he sat straight up. He then proceeded to vomit all over the back of the ambo.

Rex's next move came quickly. He stepped out of the back of the ambo, turned his head toward the busy line of lunch-time voyeurs in their cars and spewed a compact stream of partially digested potatoes across two lanes of traffic. At least a dozen windshields fell victim to the onslaught. Sean shook his head in disgust. The dead guy made a full recovery. I have never been more proud.

Rex took vomiting on the public in stride. He was the kind of guy who never got too upset about these types of things. He didn't care that the people whose cars he had hurled all over no longer loved us. The feelings that the formerly dead guy had for us more than made up for any ill will. Vomit can be rinsed off; death is much harder to scrub away.

A friend of mine summed it up best: "After 30 years of running on the people that life grinds, burns, tears and blows into little chunks, I've learned perspective. I really hope I don't have to lay my eyes on another torn-up corpse. On the other hand, it isn't as a big deal as it used to be when the dog shits on the carpet."

Chapter 12
Beacons for the Pretty & Broken

Everyone loves firemen. Firefighters and paramedics always score off the charts in public-opinion surveys. This factor plays heavily when making our initial career choice and in maintaining our happiness at the highest level. There are a lot of reasons the people have such goodwill for us, but I think the key to our success is our willingness to help people, regardless of how big or small their problem is. Another large piece of the picture is that we're nice.

For some reason, girls in particular love firemen a whole lot. I have been around women who didn't pay any attention to me when I was off duty. Less than one day and a uniform later, they became very engaging. I think it has something to do with the whole hero angle. For some odd reason, strippers seem especially drawn to us. One afternoon, we made a bunch of new friends after we put out a fire in a business next door to a gentlemen's club.

The call came in as a fire in a restaurant. We didn't see much as we pulled out of the station and headed up the road. We were almost on scene when we saw light gray smoke showing at the back of a run-down pharmacy in the middle of a decrepit 1940s-era strip mall. The end occupancies had their windows soaped over, announcing to the world they were vacant. A strip club sat next to the pharmacy.

Engine 23 arrived to the scene first, so we drove around back to investigate further. Ladder 22 found a spot on the side of the structure to spot their rig while Engine 14 parked out front on a hydrant and waited. As we pulled around back, we could see smoke coming out from the open door leading into the back of the old store. This had once been a business where one could buy back-to-school supplies for their kids, a nice bottle of perfume for their wife and get a prescription filled while enjoying a hamburger and milkshake at the soda fountain. The kindly old pharmacist would have been the owner and proprietor and would have known all of his customers. Today we call these kinds of places ancient history.

The two other firefighters and I got off the rig, threw our SCBAs on and headed inside. There wasn't that much smoke coming out of the back door, so we didn't drag an attack line off our rig. Instead, we took a 40-lb. Ansul dry chemical extinguisher and a pick-head axe with us. When we got inside, we found an out-of-control deep fat fryer burning up into the large stainless-steel vent hood. Everything was well-worn and covered with old food and grease. The fire was not very big and was still quite manageable.

A built-in extinguishing system was attached to the wall next to the cooking equipment. I pulled the pin and yanked the activation handle. Nothing happened. I yanked harder. The two large chemical tanks and all their metal plumbing fell to the floor. I still held the handle in my hand. The service tag indicated the last system tune up was performed sometime in 1965; the three firefighters standing in the burning kitchen had been 5 years old. Some of the strippers next door hadn't been born yet.

The fire continued to burn out of control, getting bigger with each passing minute. We put on our SCBA face pieces. One of us grabbed the dry chemical extinguisher, pulled the pin, and pushed the lever that dumped the high-pressure CO_2 cartridge into the powder. He then squeezed the nozzle onto the roaring grease fire. Nothing but pressurized CO_2 shot out of the tip. (This was a common problem with dry chemical extinguishers. We used them so little that the powder in them would bake during our hot summers. If you did not routinely stir the powder, it would cake up.) The big pressure blast fanned the flames and spread the fire to adjoining pieces of filthy cooking equipment. Fire filled the entire hood and rolled across the ceiling. Ladder 22 radioed us to report they had a sizable amount of fire burning out of the vent on the roof. Twenty years of grease were being liberated by fire and heat. Burning roaches fell from the flaming ductwork.

Some anonymous firefighter came in with several more of the large extinguishers. We picked them up and hurled them onto the floor to break up the powder while the fire roared in front of us. Rome probably looked a lot like this during its final days. As the fire continued to grow, it became inevitable that someone would appear with a charged hoseline.

We routinely give the public fire-safety advice. This is one reason we are so well liked. One of the safety nuggets we share with the public is never attempt to extinguish a grease fire with water. Grease and oil are lighter than water, so grease fires have a tendency to explode if you add water to the mix. Cover the pan with a lid, knock the flames down with a fire extinguisher or shovel sand onto burning grease, but for the love of God, do not put water on a grease fire.

The firefighter who showed up with the hoseline applied water to the burning deep-fat fryer, directing a straight stream of 125 gallons of water per minute directly into the vat of burning grease. When the column of water blasted into the burning oil, the liquid explosively distributed fire all over the old kitchen's walls and floor. The oil did what oil does. Since it is lighter than water, it rose to the surface of the 100 or so gallons of water that covered the floor and quickly re-ignited, moving toward us like lava flow. The grease got all over our big rubber firefighting booties. This caused the group of us to slip and fall. The guy with the line was one of the first to slip. When he fell, the line slipped out of his hands and began thrashing back and forth, like a rabid python. The group of us frantically kicked away from the flow of burning oil as the open hoseline beat the shit out of us. One of the firefighters from the ladder came in, grabbed one of the extinguishers and put out the fire in about 15 seconds. He looked at the group of us like we had just lost our minds. We collected ourselves, shut down the errant hoseline then headed out the back door. We were taking our turnouts off when a group of employees from next door came over to see what the ruckus was about. The girls said they were eternally grateful that we saved their place of employment. We invited them to the station for dinner. They didn't get off work until well after dinner but said they would try to make it.

We finished the dinner dishes and were sitting around polishing the brass when five of our new stripper friends showed up. They looked a lot different wearing normal civilian attire, like secretaries, waitresses or cheerleaders. The girls had come by to take us up on our offer for dinner. We had more than 20 lbs. of leftovers, so feeding the svelte dancers would be no problem. They ate enough food to feed a school of homeless children. After the girls finished their dinner, we all gathered around to discuss the highs and lows of their job. All of this was leading up to an after-dinner performance. Before the strippers could practice their chosen profession one of the captains thanked them for stopping by and wished them a good evening. Prior to the girls leaving, they traded contact information with some members of the station. This led to several very satisfying short-term relationships.

☠☠☠☠☠☠☠☠

Another group that is inexplicably drawn to us is the mentally retarded. According to the most senior members of our department, it has always been this way. It was so prevalent at one time, the mentally challenged formed their own organization. It was called the 2-11 Club, but they later changed their name to the Fire Buffs. Someone procured an old, worn-out snack-wagon truck, painted it red and parked it at a fire station. Only members of the club who had a driver's license (many of them didn't drive) could drive the rig to fires. The truck would set up at the scene and provide drinking water and very old granola bars for tired firefighters. The snack wagon was pulled out of service when one of the unlicensed savants took it out for a two-hour drive with lights and sirens.

Some of these mentally challenged "associates" are legendary, and each generation has theirs. It is inherent to our service and cannot be stopped. It's like little green apples in the summertime. My muse is a man who goes by the name of Mr. Blueberry. Blueberry is half Navajo and half Chippewa Indian—we call him a Chippaho. He has the intelligence and emotional maturity of an 8-year-old. His head is seriously misshapen; his palate never formed properly, so he is left with a very crooked grin. Standing almost 6 feet tall and weighing 220 lbs., Blueberry is a very large man.

Blueberry loves gadgets and clothing. He likes to layer plaids over stripes and then cap them with a V-neck vest adorned with a giant embroidered kitty cat. He wears white patent-leather zip-up boots and a wide belt with a big brass buckle that reads "BLUE." His shirt pockets are stuffed with writing devices. His vest sports an array of electronic gadgets that tell you the time and temperature, to have a nice day, and make various animal noises. Other battery-powered units blink, chirp, light up and create siren noises. One of my friends says he looks like a Ferris wheel coming at you. Blueberry scares people who don't know him.

We hooked up with Blueberry when I was working at Station 66. Blueberry wandered into the station one day and ended up staying for the next 10 years. We sort of adopted each other. He would clean the station, which is something we hated doing, and we would pay him a few bucks and feed him lunch and dinner. The arrangement was mutually beneficial.

Blueberry lived across the street from our station in an apartment that can best be described as third world. He had electricity and running water, but nothing else. I have been in quite a few shit holes over the course of my career, but few of them match the squalor of Blueberry's pad. Not even the swamp cooler worked properly. The interior temperature at Blueberry's place hovered at 100 degrees for months at a time, so he basically lived at our station during the summer. This was a no-brainer, even for someone with diminished faculties. The inside of our station rarely got warmer than 70 degrees. Blueberry's hygiene improved remarkably when he started hanging out with us. We forced him to bathe and see the dentist. He started doing his laundry in our station washer and dryer. The payoff for us was Blueberry's constant presence. Our station was in a very seedy part of town and had been burglarized on more than one occasion. It often sat vacant while we responded on calls. Blueberry kept an eye on things and ran off any riff-raff.

Eventually, Blueberry took over as our receptionist. The phone in a fire station can ring dozens of times an hour. We are one of the most social groups of people on the planet. (Imagine 10 teenage girls living together and sharing four phone lines.) Everyone wants to talk on the phone, but no one wants to answer it except Blueberry. Most big companies would not allow Mr. Blueberry anywhere near their telephone. Because the bones of his mouth, face and head had never fused properly, his speech is difficult to understand. Plus, he's mentally retarded. Communicating with him is a lot like talking with a stubborn and drunk 8-year-old. Fortunately, most people calling our station were other firefighters who knew Blueberry. In fact, Blueberry usually spent more time talking to the people who called than we did.

One day, my boss stopped by to ask me about our newfound friend. Not everyone was thrilled about having our fire stations serve as hangouts for the mentally impeded. They argued fire stations were places for official city business, not halfway houses for lunatics. Most of the people assigned to my station fell into the "lunatic" category. Although my boss wasn't too wild about Blueberry, the only concession he asked for was that Blueberry quit answering the phone.

I understood the concern some members had about our special friend. Surely we had enough to manage without throwing Blueberry into the mix. After all, the Blueberries of the world come complete with their own unique set of problems. Still, we believed that since we routinely made life-and-death decisions in lethal situations, we should be able to manage Mr. Blueberry.

After my boss left, I sat Blueberry down and told him he was doing a great job around the station. So good in fact that we were going to raise his pay from $4 to $5 per hour. As another bonus, he didn't have to answer the phone any more.

"Blueberry, we will start to answer the phones again. You do too much around here as it is."

"Oh no you don't. You ain't gonna know that I like to answer them phones."

"No, Blueberry. We will answer the phone."

"You listen here now, Mr. Nick. Blueberry will answer the phone. Don't worry your pretty little head about it."

Blueberry grabbed me around the neck and started to stroke my hair. I was not getting through to him. I tried a different approach.

"Blueberry, the chief just left here. Evidently, the mayor just called him and wanted to know why the firefighters at Station 66 had you answer our phone. He told him if you kept answering the station phones, the city would let one of us go. So there you have it, Blueberry: If you don't stop answering the phones, the city is going to fire me."

I blame myself for what happened next.

The morning of the next shift I was sitting at the kitchen table when the phone rang. I answered it, "Station 66, Brunacini speaking. How may I help you?"

A voice screamed at me, "Is that man there?"

"What man?"

Still yelling, "Don't toy with me. You know what man!"

"No, I don't. I don't now what man you're looking for or who you are, and if you don't tell me, I will hang up on you by the hair of my chinny chin chin."

"This is your battalion chief, and I demand to know if that man is there! He called my office and left a threatening message."

"This doesn't sound like you, Bob. What man left a threatening message?"

"This isn't Bob, it is Chief Floyd. I'm working for Bob today, and I'm ordering you to escort that man out of the station if he is there. I have called the police, and they are on their way over to the station as we speak."

"What man?"

"That retarded Indian you guys keep as a house boy. He just called my office and left threatening remarks on our answering machine. I will be there in 5 minutes. I had better not see that man in the station when I get there."

After I hung up, I turned and asked Blueberry if he had called my battalion chief.

"That guy got no right to fire you, no he don't. That's a question for you, Nick Bruci."

"Did you leave a message, Blueberry?"

Blueberry flashed me a crooked smile, gave me a big bear hug, and told me he loved my face.

I told Blueberry he had to go home for a few minutes while we got things straightened out. He was not happy about leaving. I walked him across the street and told him I would be back in half an hour.

When I got back to the station, Jimmy — one of our internal auditors — and Chief Floyd awaited me. The chief was still yelling.

"I will not allow this to continue. I want to press charges against that man. He left threatening remarks on our battalion answering machine." As if it were some type of holy recording device...

Jake the engineer asked Chief Floyd if he had the answering machine so he could prove his allegations against Blueberry. Chief Floyd pulled the answering machine out of his briefcase, plugged it in and pushed the play button.

"You got no right to fire Nick Bruci. He is my friend and those guys at Station 66 need Blueberry to answer their phone because they are always going on the calls and Blueberry writes their messages on the board. You need to mind your own damn business and get the devil out of your head. I pray to Jesus, Lord Jesus, please get the devil out of the mayor's head and help him to mind his own damn business." This went on for the entire 20-minute tape.

Jake, Jimmy the auditor and I were all laughing so hard we were hyper-ventilating. At one point in the tape, Blueberry said he was going to "throw a prayer on the chief." Chief Floyd mistook this for "throw some pain on you." Blueberry wasn't easy to understand.

Jimmy asked me if he could meet Blueberry. As we walked across the street to Blueberry's hovel, the auditor made a remark about that "ignorant moron." I had had my fill of this morning's nonsense. I told Jimmy Blueberry was just trying to protect us and was angry because he thought I was going to be fired. Before I could get too carried away, Jimmy stopped me and said, "I was talking about Chief Floyd."

We told Blueberry we were taking care of the situation, but he couldn't come around until lunch. After seeing Blueberry's apartment, the auditor asked if there wasn't somewhere else he could stay for a little while. Blueberry had become quite renowned within our organization and had more than a dozen stations he frequented; he was a regular prince of the city. Blueberry loaded up the four bags of shit he dragged everywhere and headed down the road mumbling, "Blueberry's going to catch the 14 redline to the Number 7 and get off at the Van Buren stop to see Mr. Jackson." Blueberry was taking the bus to go and hang with the shift commander, Jackson James.

When I walked back into the station, Jake was screaming. "I was quiet and listened to you, now you're going to give me the same consideration, you fucking asshole."

As I came around the corner, I saw he had been addressing Chief Floyd, who became very irate, collecting his tape recorder and heading for the door. He turned before he left and said, "You haven't heard the last of this. The inmates will not be running the asylum on my watch."

When Floyd was finally gone, I called Jackson and told him what had happened. Jackson had a picture of Blueberry sitting on his desk. It had been taken at the James family reunion. Blueberry, looking like a Ferris wheel, was holding the end of a slinky. The other end of the coiled steel toy trailed 20 feet behind him before disappearing off of the edge of the picture. Jackson said not to worry about anything, he would keep his eye out for Blueberry.

Floyd was so infuriated about Jake's tirade that he called the shift commander when he got back to his office. Chief Floyd's day got worse when Mr. Blueberry answered Jackson James' phone.

One of the largest publishing houses in the world is the German company Bertelsmann. Every year, they choose a particular industry and give an award to whomever they deem to be the best of the class. In 1994, they were researching which was the best-run city in the world. The Phoenix Fire Department contacted the publishing giant and threw our city's hat into the ring. A few months later, they chose Phoenix as the winner. Many other cities were jealous, which added to our civic pride. During the week the German presentation committee was in town, they had arranged to have lunch at one of our downtown fire stations. The fire chief, union president, mayor, city manager and other important city people would be there. Lunch was going to be a big deal.

The morning of the big lunch, I was working out in the gym located in front of our station. Looking out of the windows, I could see Luke, one of the engineers on the ladder, helping Blueberry fix the cooler that was attached to the side of his shack. They had fooled around with the rusted unit for 15 minutes and were on their way back. They stood on the sidewalk waiting to cross the busy street. It was quite a contrast. Luke looked like a male model. Blueberry looked like a train wreck—a circus train. He was carrying his four totes full of shit. From where I was standing, I could see both Blueberry and Luke and about 10 feet of road on either side of them. Blueberry suddenly reached into one of his bags and pulled out a full-size stop sign that he had attached to a crude wooden handle. On one side of the sign, under the word "stop," he had spelled out the word "car" in big, white letters. On the other side of the sign, he had written the word "bus." Blueberry looked at the sign, flipped it to the "stop bus" side and stepped into the middle of the street. He turned to face oncoming traffic and held the sign high in front of him. He executed this maneuver in one smooth motion, like a striking cobra. Before Luke could react, Blueberry was standing in the middle of the street like the Statue of Liberty on peyote. In the time it took me to register what he was doing, I heard the sound of locked-up tires skidding across blacktop.

Most people will only see a very long, large and out-of-control city bus in an action movie. For the most part, bus drivers do not put their rigs into

132

sideways 40 mph skids. I sat on a weight bench watching the fast-moving bus trying to trade ends. It was almost perpendicular to the sidewalk when it reached the point where I had last seen Blueberry defiantly standing, with his outstretched sign, commanding the bus to stop. Smoke rose from the tires as the bus continued down the road at speeds fast enough to kill and maim. In the blink of an eye, the bus had veered past Blueberry and Luke. Blueberry was safely standing on the sidewalk, having been rescued from certain death by Luke. The bus driver brought the front end of the bus back around and continued down the road without stopping. He had a schedule to keep. We had the best-driven buses in the best-run city in the world. We truly did deserve an award.

Mr. Blueberry attended the big awards lunch. The fire chief and the union president both sat at their table and watched as Blueberry made his way through the chow line. He was sandwiched between the city manager and one of the German industrialists. The chief and union prez looked at each other and quietly accused me of giving Blueberry bus fare and sending him to the gala affair. Nothing I say will ever change their minds, so I will no longer attempt. Both of them said no one even noticed him; he fit right in.

Firefighter's biggest fans are children. Kids love firefighters and the big, red trucks we ride. Each one of us has several defining career moments; a large group of children provided me with one of mine.

It was the Friday before Easter, and we were scheduled to deliver the Easter bunny to the ghetto school across the street from Station 55. It was just before lunch when the bunny showed up at the station. He was a small guy. While he was putting on his rabbit suit, one of the teachers came to the station to give us directions. A gate off of the main road would be open, and we could drive across the playground and deliver the pink bunny to the assembled group of 1st graders.

The rabbit came out of the bathroom, put on his head, and we all headed out to our beloved fire engine. We only had four seats, so I rode on the tailboard. My captain engaged the lights and sirens as we pulled out of the station and turned onto the road. A smiling custodian waved us into the school's open gate. Our engineer slowed down as he drove over the curb and sidewalk and onto the grassy playground. Hundreds of kids were lined up a few hundred feet in the distance. When they caught sight of the rig, they went crazy. The teachers were moving up and down the assembled line trying to keep order. They looked like riot cops. I was standing on the tailboard, looking down at the grass we were driving over. It was wet, and the tires were beginning to sink. The pink bunny stood up and began waving at the kids, provoking a greater frenzy. Now the rear tires were beginning to spin, and the truck was hesitating. We were sinking. Our rig was still more than a hundred feet away from the waiting kiddies, and we were not going to move another inch. I jumped off the tailboard and walked over to the dual rear tires. The engineer floored the gas in one final effort to free the rig, throwing a 30-foot mud rooster tail into the air. The Easter bunny turned its matted fur head toward me and shrugged his shoulders. The kids were now uncontrollable. I waved my arms at them and shouted at the rabbit to scamper to his adoring audience. Mr. Rabbit hopped off the rig and raced toward the children. The teachers could no longer contain the rabid kids. They broke free and headed straight at the fake bunny. It was quite a sight to behold. Easter bunny and the kids were far enough apart that they reached full speed a hundred feet away from one another. Kids were pumping their little legs and arms; the rabbit's head bobbed back and forth as candy bounced from his basket. When they finally collided, a very strange thing happened. The bunny hit the ground, but the kids kept coming. They couldn't give a shit about the rabbit. They wanted to play on the fire truck.

Chapter 13
The Last One

A career in the fire service has little in common with other occupations. Riding on big red trucks, running into burning buildings and occasionally bringing the dead back to life make our jobs unique, but it's our work schedule and career span that truly set us apart. The prevailing fire-service shift is 24 hours on, 48 hours off. Firefighters spend a full third of their working lives living together, and careers lasting more than 30 years are common. Each shift includes a sleepover with your workmates. The only other work places that come equipped with beds are military barracks and brothels.

Spending 24 hours together every three days for 30-plus years builds deep bonds. Belonging to such a close-knit group creates a unique culture. We have our own customs and identity. We have our own sense of humor. Not everyone understands us. In reality, the group of firefighters assigned to a station operates more like a family than a collection of coworkers. It says so right on the side of all Phoenix Fire Department apparatus: "Our family helping your family." (It has been suggested we should revise our motto to more accurately say, "Our dysfunctional family helping your dysfunctional family.")

Within the collective family of any good-size fire department, each station becomes a unique sub-unit. Within these sub-units, the three individual shifts are a lot like cousin families to each other. If you were to graphically illustrate the family tree of the average American fire department, it would look a lot like the charts and diagrams the FBI uses to understand Mafia family ties.

A life in the fire-service family can have a lasting impact on one's definition of normal human interaction. Our skewed social practices often run counter to the current plague of political correctness that overwhelms the rest of society. This is because families are not politically correct with one another.

Some families are caring and nurturing. Others are caustic and dysfunctional. Like many families, members of the fire-department family will risk their lives for one another. They will lie for each other, enable each other and even loan money to one another. Some members of the organization become so close, they end up trading spouses.

Within the closed confines of a fire station, every member has a unique reputation. We know each other's strengths and weaknesses. We also quickly figure out how to push one another's buttons. Fire-station life is not for the weak.

During the mid-1990s, the Phoenix Fire Department had a never-ending stream of out-of-town visitors. People came from all over the world to study us. During this era, the city of Phoenix reached out to a group of budding new capitalists from Communist China. I'm sure the members of our local chamber of commerce could barely contain themselves over the prospect of cheap Chinese labor. We sent over a few thousand pounds of citrus, and in return we received indentured Chinese slave girls to work in our jalapeño-jelly and cactus-candy factories. It was a heady time of building international relationships through commerce.

I was about to promote from captain to battalion chief and was working as a chief's aide in downtown Phoenix when the first coalition of unsuspecting Chinese dignitaries came to our Wild West town. I received a phone call from one of the higher ups, informing us the mayor of our Chinese sister city would be riding with us after lunch. My direct supervisor was a man whose mother had named him Jackson. Upon receiving the news, Chief Jackson James simply rolled his eyes and threw me an ambiguous shrug of his shoulders. Jackson never got too excited about visiting with out-of-town Communist capitalists. He was a cool cat.

The first time I had the pleasure of working with Jackson, he was a paramedic-captain on C Shift. I was working an overtime shift on an ambulance. We had been special called by Jackson's engine company for a gunshot wound. The patient had just left his day job as a rocket scientist and drove over to his girlfriend's house. After exiting his car, he jammed a cocked and loaded semi-automatic handgun into the front of his pants. The gun went off, blowing most of his left testicle through his right thigh. We got the guy into the back of the ambo and cut off his blood-soaked pants. Jackson looked at the wound and said to the gunslinger, "That looks like it really hurts." The guy looked down at his maimed manhood and passed out. I asked Jackson if he learned his bedside manners in paramedic school.

After lunch, the mayor showed up with her lovely and exotic female interpreter. I was surprised to see the mayor was a woman. Jackson had cultural leanings toward beautiful women. He took charge and showed the pair around our office. The mayor and her grad-student tour guide talked nonstop. The guide asked a never-ending stream of questions. A lot of "What is this for?" "Why do you do that? "How come?" kind of stuff. We were trying to help the pair make sense of Western firefighting when we got a call. Our Asian friends jumped at the opportunity to go.

136

The four of us were responding to a reported house fire. The two visitors carried on in the backseat like a pair of schoolgirls on a roller coaster. The first engine arrived to the scene and reported a trash fire that they could handle; they canceled the rest of the assignment. As we drove back to our office, the interpreter asked, "The Mayor would like to know why it takes two of you to manage a fire." Jackson started to give the textbook management answer: "While the aide drives, I listen to the radio traffic, pull the pre-plans and start filling out a tactical worksheet." The interpreter regurgitated this back in Chinese. I could see the mayor in the rearview mirror. She looked confused by the answer. I told Jackson to quit lying to our out-of-town guests. I turned to face the mayor and told her, "It can be very stressful to run a fire; it is my job to bathe Chief Jackson-son when we get back to the station." Jackson turned around and quickly quipped back to the interpreter, "That's a lie. Please do not repeat this to the mayor." It only took the interpreter a nanosecond to figure out that she was going to tell the mayor I was Jackson's 250-lb. love slave. She began speaking a language that neither Jackson nor I understood, but we knew exactly what she was saying. The color drained from the mayor's face as her young tour guide told her about American firefighter man love. I screamed at Jackson, "Don't try to hide the truth, mister! I will not let you fool our most honored guest." I turned to the interpreter and said, "I also get him tea and rub soothing emollients into his bunions. I am his humble servant." Jackson screamed at me, "Nick, shut up, goddamn it! These people don't know you're brain damaged. They believe everything you're telling them." I cowered when shouting in reply, "I am so sorry Jackson-son! Please do not beat me with your big man stick when we get back. I cannot live with your disapproval. I will be a good bitch boy from now on." It sounded like someone was blowtorching a pair of bag pipers in the backseat. The interpreter was pointing back and forth between Jackson and me as she spoke very excitedly to the mayor. The mayor was making doorbell noises and shaking her head in disbelief. When we got back to quarters, the mayor and her interpreter quickly exited the vehicle and briskly left the station without bowing.

There was no fallout from our higher-ups concerning the potentially devastating international event. Maybe the mayor and her language companion chalked it up to the lawlessness of the American West. Being an ambassador isn't all fun and games. Problems often do arise when two cultures converge.

☠☠☠☠☠☠☠☠

Managing a fire station is one of life's great pleasures. Delivering a full menu of emergency services to the community is exciting work. The other joy of supervising a fire station is getting to quarterback the fascinating set of idiosyncratic firefighters that make up a fire company.

For the better part of a decade, I was assigned to a busy ladder company. During my tenure at Ladder 66, I worked with Jake and Luke, a pair of the most capable engineers on our department. Luke was the most senior member of our station. He had been sent to Station 66 early in his career as punishment for having long hair. Back in the early 1970s, one proved their love and submission to the family by having fire-department-approved hair. The chief ordered Luke to get a regular "man's hair cut." This was during an era when young people across the country were bucking authority. When Luke refused to comply, he was reassigned to a fire station loaded with organizational castaways. These places always offer the most vibrant of fire department sub-families. Luke quickly discovered he enjoyed working with the misfits at Station 66. He kept his hair long and spent 37 of his 38 fire-department years working at 66.

Luke was no stranger to fighting the system prior to joining the Phoenix Fire Department. During the Vietnam War, Luke found himself with a draft lottery number that assured him a job in the armed forces. He decided his time serving our country would be better spent guarding our nation's coastlines against an invasion from the North Vietnamese Navy, so he joined the United States Coast Guard. Luke breezed through his training with flying colors. After finishing his training program, he was assigned to a ship stationed in the Hawaiian Islands. Luke thought he had died and gone to heaven until his ship received orders to go on an ice-breaking mission on the Alaskan coast. Reasoning that a three-month stay in a Hawaiian brig was better than the frozen Arctic seas, Luke chose to go AWOL and remain in his island paradise. His plans fell apart when the MPs arrested him and transported him to a waiting helicopter. Luke asked his captors where they were taking him. One of the MPs told him, "You'll be serving your three months in the brig aboard your ship. As I remember, the brig on a Coast Guard ice cutter is in the bottom of the ship. You should have brought a jacket with you."

Luke was the only member of our station's B Shift family who served in the military. This changed when a probationary firefighter named Roger cycled through our station. Roger joined the Phoenix Fire Department family at the age of 42. He was a couple months older than Luke, who had already put in more than 20 years and was eligible for retirement. Roger began his first day with the standard overview I gave to all the new firefighters. During our meet and greet, I learned that Roger had decided to pursue a career he could enjoy after spending the last two decades running his own business. During the course of our conversation, we started talking about each member's tenure. I rattled off how much time each member had been on the job, ending with Luke, who was hired in 1969. Roger had a far-away look in his eye as he thought out loud, "In 1969, I was sloshing through my first tour of Vietnam." All I could offer was, "Wow. In 1969, I was in the 5th grade."

I ended our meeting by explaining what we expect of our junior members. During a firefighter's probationary year, we expect them to make the coffee, put up and take down the flag, answer the phone and be the point person for station cleaning. Basically anytime any member of the station is working, the new guy (or gal) should lend a hand. The routine was designed to let the new member know they were at the bottom of the totem pole. In addition to being a right of passage, the booter routine taught the concept of teamwork. I told Roger not to worry about all the typical probie bullshit. He would do just fine if he lent a hand to whoever was working—basically to act like any other member. Since Roger was the oldest member of the station, I told him he should catch a power nap after lunch because he could plan on going on a minimum of two calls after midnight.

A couple shifts after Roger joined our Station-66 family, Jesse, one of the paramedics assigned to Engine 66, came into my office after lunch. He woke me from the sweet release of my slumber. My first thought was that we had a call. Sometimes when you wake up prematurely, you have the feeling that the village vagrants stoned you during your sleepy time. After I came to, I rolled out of my rack and started to head out to the rig. Before I got out of my office, Jesse told me there was something he wanted to show me. I told him it would have to wait until I got back from the call he had just woken me up for. Jesse looked at me like I was retarded and told me we didn't have a call (in fact, I think he told me the ladder never went on calls); he had woken me to show me something. It's usually bad news when someone has to wake up one of the captains to show them "something." My next thought was that some type of catastrophe had befallen the station. Maybe one of the firefighters had inadvertently blown themselves up with one of their homemade potato cannons.

I followed Jesse out of my office and into the station dorm. It was dark and cold. The dorm was a long rectangular room with no interior walls that held 10 twin beds. Each bed was separated from the next by a set of lockers. The inhabitants of the dorm had run a length of pipe from one locker to the next and attached black shower curtains to the pipe. This provided a modicum of privacy. It also gave the dorm the look and feel of an opium den. Jesse led me to the bunk closest to the door that led out to the apparatus bay and pointed at the closed curtain and said, "You want to see what Roger is doing?" I didn't know what to expect. Twisted thoughts ran through my mind. Was Roger caught up in some type of demonic ritual involving half-dead poultry and severed goat heads, or had he killed himself from the overdose of joy he was experiencing from working with such swell fellows? I tenuously poked my head between the curtains and saw that he was lying in his bed with his arms folded over his chest and his ball cap pulled over his eyes. He was fully dressed and appeared to be resting more than actually sleeping—an activity I was fully immersed in about 2 minutes ago. I pulled my head from between the curtains, turned back to Jesse and asked, "Big deal. What did you want to show me?" Jesse gave me another look that implied I was very dense and said, "You don't find anything odd with this picture? This is the booter's first station. He hasn't worked here for two weeks, and he's already taking a nap. I think there is something seriously wrong with that. If I had done something that ballsy when I was on probation, I would have been written up, maybe even fired."

Before I could respond to this mind-numbing encounter, someone had affectionately screamed from behind a black shower curtain, "Jesse, you're a stupid Mexican. Shut the fuck up. We're trying to sleep, you rude, lame-ass, thumb-dick fuck-head." Jesse appeared to partially deflate at this barrage of harsh criticism.

We are all prisoners of our brains. My brain constantly assaults me with images, distortions and other noise that can be quite distracting. I flashed back to an event that occurred a month earlier. I had been working on the engine with Jesse. After lunch, we were going on a call when Jesse's partner was overtaken with laughter. I turned around to find out what was so funny. Jesse had taken off all his clothes and was sitting buck-ass naked, belted in his seat. When we made eye contact, he blew me a kiss while he gingerly rubbed his little mi hijo. I smiled and imagined how proud Jesse's deeply Catholic parents would be if they could see their little boy rolling down the road, wrapped in gleaming red and chrome, naked as the moment he popped out of his brown mama.

Standing in the cold, dark dorm, two thoughts flooded my mind. First, the guy who had just screamed disparaging remarks to Jesse was Engine 66's driver, John, and he knew Jesse's dick size from his habit of riding on the truck naked. Second, Jesse had screwed-up priorities. He found nothing wrong with riding nude inside a 16-ton vehicle that called attention to itself with an array of flashing lights and a set of sirens that could raise the dead. He did find it a personal affront that a junior member of the organization had the unmitigated gall to take a nap before their one-year anniversary. The madness was more profound when one considers Jesse and I were the only two members of the 10-member station who were currently in a vertical position. I told Jesse he really shouldn't worry about Roger taking a nap. "Roger is old enough to be your father. Besides, if he doesn't get enough rest, his bones will weaken. If he falls, he could break a hip. And more to the point, Jesse, if you want to manage the new guys, take the captain's test." Jesse looked at me reflectively, nothing like he had looked at me when he was going on a call naked. His thought-out reply was cut short when John shouted from his plastic cocoon, "Yeah Jesse, take the captain's so we can call you Captain Dick."

I spoke up and said, "That's enough." John replied, "I'm sorry, Jesse. I meant to say we could call you Captain Thumb Dick."

The morning of our next shift we were sitting around the dayroom table, drinking coffee and collecting money for chow when John assailed Jesse for interrupting last shift's nappy time. Their bickering was interrupted when one of the group told Roger to quit mopping and sit down. Roger complied with the request and sat at the end of the table with a big squinty smile. The previous shift had been too busy for the whole station to carry on the standard "what did you do before you came here" discussion. Roger told us about his wife and kids, where he lived, and what he did for a living before joining our cult. Someone pointed out that Roger was one of the oldest firefighters our department ever hired and asked what caused him to join the fire department so late in life.

Roger told us that after he came home from the war, he started a series of businesses. "I always seemed to be able to make money, but it was just work. Five years ago, I became friends with a couple of Phoenix firefighters. The way they talked about their careers here was the way I felt about the unit I served with in Vietnam. It was the first time I've heard someone describe their coworkers like that since I left Southeast Asia." One of us asked, "Let me get this straight. You joined the fire department because you miss Vietnam?' Roger replied, "I miss the relationships I had when I was over there. My unit had a bond with one another that doesn't exist in the working world that I've been a part of since I left the service.

They were the last group of people I worked with that I really trusted."
Now the group wanted to know more about Roger's Vietnam experience.
"I was a grunt and did two tours," Roger explained. "The government made
me leave after my second tour was up. I left as a sergeant. My unit went
all over the country. We had really good officers. They always made sure
we had the upper hand when we went into battle. If you did what you were
trained to do, you didn't have much trouble. We gave a whole bunch better
than we got. The North Vietnamese had a bounty on our unit. None of us
was too concerned with it because we were pretty good at what we did."
Roger had everyone's complete attention. "You guys have no idea how
lucky you are," he continued. "Organizations like this are a rarity." Before
we broke from our morning camaraderie, John turned to Jesse and asked,
"You got anymore to say about Roger's nap?"

☠☠☠☠☠☠☠☠

I spent a decade working at Station 66. The relationships we B-shifters
formed during those times are best described as family ties. Every third
day, the group of us would leave our mortgages, parents, wives and children and go off to spend a 24-hour shift with our other family. Each
member of the station had a urine-stained driver's seat from pissing a big
wet spot of excitement every time we drove to work. It is an odd phenomenon: We routinely respond to incidents of brutal human tragedy, but we
seem inoculated from it. In fact, it is even stranger than that. We are a group
who enjoys responding to calls of death and destruction, but I've never met
a firefighter who enjoyed watching people suffer. It is an intriguing
dichotomy that we perform at our best when the world is vaporizing around
us. Roger's 20-year-old analysis was correct. The two most powerful
coping mechanisms a fire company has to protect itself from the nasty
horrors it responds to is being well trained and being part of a tight-knit
group that trusts and depends on one another.

It's tempting to end this essay with the sentimental thought that I'm the
luckiest man alive for stumbling into a career in the fire service, but that
would be half-baked conjecture. What if I had actually applied myself in
school? I could have ended up winning the Nobel Prize for discovering a
source of clean, cheap power. Suppose my interest took a slight turn toward
a love of ministering? I could be the proud billionaire owner of my very
own Jesus Christ Amusement Park and Redemption Center. On the other
hand, my life could have just as easily nudged down a different fork in the

road, and I could be serving 25 to life in a maximum-security prison for any number of reasons.

They say when you help your fellow man, you get back more than you give. Firefighters are very fortunate in this regard. We live careers that allow us to make a difference in other people's lives. We are the only government agency that shows up at your door within minutes of being called. Our only question is, "How can we help?' If the problem is life-threatening, we will immediately place our bodies between the customer and the incident hazard. Our help and assistance require no bureaucratic forms, permits, inspections or permission from the body politic. Just as amazing, once we solve the problem, we clean up our mess and go away. (I defy you to name another government agency that allows the customer to return to a normal life without continued government intervention and control.) One of the beauties of our job is you never know what you're going to encounter when the lights come on and the bell rings. Every call is its own story.

The most powerful form of cultural communication within the fire service is the war story. We use stories to record and share our past. Every firefighter who ever lived has sat around a kitchen table, the back step of a fire engine, or in the front yard of some burned-out building recanting the details of a call. These conversations always end with the words, "Someone's got to write this shit down." Despite the fact that I didn't invent a new form of power or open an amusement park for God (or go to prison), I am fortunate to have the career I've had and the opportunity to write some of it down.

I began my fire-service adventure in 1980. It has been more fun than rock-and-roll. For 26 years, I couldn't have asked for a better organization to work for than the Phoenix Fire Department. In every sense of the word, the Phoenix Fire Department was my family. Fire-service professionals from around the world have referred to it as Camelot. My career has allowed me to plunge my head under life's skirt and gaze at her wonders. The view has been grossly fascinating, hysterically funny, heartbreakingly sad, occasionally heroic and richly rewarding. It also smells a lot like fish.

CPSIA information can be obtained at www.ICGtesting.com
Printed in the USA
LVOW071645141112

307335LV00024B/105/P